The World Tree

Gathering for the Shift of 2012–2013

The World Tree
Gathering for the Shift of 2012–2013

by William Gaspar and Jose Jaramillo

℮

Eloquent Books
Durham, Connecticut

© 2010 by Dr. William Gaspar and Dr. José A. Jaramillo
All rights reserved. First edition 2010.

No part of this book may be reproduced or transmitted in any form or by any means, graphic, electronic, or mechanical, including photocopying, recording, taping, or by any information storage retrieval system, without the permission, in writing, from the publisher.

Eloquent Books
An imprint of Strategic Book Group
P. O. Box 333
Durham, CT 06422
http://www.StrategicBookGroup.com

ISBN: 978-1-60911-321-6

Book Design by Julius Kiskis

Printed in the United States of America
18 17 16 15 14 13 12 11 10   1 2 3 4 5

Contents

FOREWORD vii

INTRODUCTION ix

1. The Sacred Feminine VA and the Galactic Heart 1

2. The Rock, the Red Hero and the Precession of the Ice Ages 31

3. From the Iron Rod of the Rock the Hawk Flies South 71

4. Venus, Darkness and Beheading 92

5. The Chariot, the Ocean, and the Sacred Twins 121

6. The Cosmic Language 148

7. Ancient Mayan, Hindu, and Egyptian Roots 168

8. The Crocodile Dragon of Mythology 189

9. The Great Fish Whale 212

10. The Lost Codices of Pakal 228

11. Decoding the Tomb of Lord Pakal 251

12. The Last Twelve Hours of the Hero's Journey 284

13. The Real Cosmic Secrets and The Da Vinci Code 306

BIBLIOGRAPHY 327

Foreword

Is the world going to end in 2012? That is the question the authors try to answer in this book, *The World Tree: Gathering for the Shift of 2012–2013*. They found overwhelming evidence in the coded language of mythology and science from the ancients. The Mayans, the Egyptians, and other ancient cultures applied similar symbols in their celestial cosmology to explain the earth change scenarios of the next shift of the ages.

Historical and literary sources lead to the cracking of myths . . . After decoding complex ciphers of celestial activity they bounced on astonishing astronomical data, the remains of an age-old astronomy that foretells a coming cataclysm in 2012!

For those who are interested in knowing what is going to happen at the end of the current cycle of the sun, this book reveals an ancient secret about how the next shift of the ages will take place on Earth. It is clear that the authors of this fascinating book are trying to show us a real cosmic model of how all these things are going to happen. What were the Egyptians trying to tell us? What about the message of the Hopis and the Mayans, who call our current cycle the Fourth Sun, or the fourth cycle of civilization? Even with overwhelming details coded in stone to last the tests of time, the scientific fields still claim ignorance about what awaits. *Scary* is an understatement of their work. This fascinating book puts science on its head and has to be taken very seriously by the reader.

—Patrick Geryl, *The Orion Prophecy*

This is an amazing new understanding of our world's relationship to cosmic time cycles and their effects upon man and the evolution of our path to the coming events of 2013. It is a left-field approach to a little understood science that is crying out for greater attention. It brings forth the place where the stars play a role that is affecting life on earth and has the Cygnus constellation as a key to this age-old mystery.

—Andrew Collins, *The Cygnus Mystery*

Twenty years ago I had a close encounter with a "deceased" former colleague, a "man of light" literally now. That shocking event turned my mind inside out and my life upside down. For at that time there were no Owner's Manuals for after-death communications or much research into near-death experience correlated with future visions of coming earth changes.

Ten years ago I came in contact with an extraordinary scholar, William Gaspar, M.D., not only a physician, but a metaphysician. Our encounter had begun an idea exchange to connect the dots from the star constellations to creation mythologies. He and I sensed our heartfelt kinship immediately and urgent common goal to prepare the population of this war torn world for the shake, rattle and roll. Thus, William Gaspar and Jose Jaramillo wrote The World Tree for the era of 2012-2013. I recommend it without reservation.

—John Jay Harper,
Author of *Tranceformers: Shamans of the 21st Century*

Introduction

Throughout known human history—as long as we can remember, for every tribe, culture, and religion—a tall, sky-scraping **Magical World Tree** symbolized sacred **life-giving** and maintaining forces. This World Tree was Rooted Deep in the heart of the Earth and towered high in the sky to reach the heavens. This World Tree was called Skambha by the Hindus and Sampo by the Finnish people. To the barbarian Vikings, this World Tree was known as Yggdrasil, the World Ash Tree. In the ancient cultures of the Babylonians, Hebrews, and Egyptians, it appeared as the towering Cedar of Lebanon. Some Native Americans, such as the Lakota and Dakota tribes, celebrated it as the Sundance Tree, where the Sacred Tree was tied to the function of the Sun. The far-seeing and enigmatic star priests of the Mayans called it the Calabash Tree and had an elaborate cosmological tale about the two founding brothers of the Mayans who went through some hair-raising and challenging ball games to access the Tree.

These ancient star priests had a deeply rooted belief system based on eternity of spirit and the changing cycles of time. The shamans, medicine men, priests, imams and rabbis readily understood, deep down in their divine spirit, that in these cycles there is no end or beginning; these cycles are the **never-ending cycle of life**. Thus, these star priests and priestesses had to be well versed in the observation of the movement of the stars. They made advanced systems for counting days and they wrote down their knowledge of cosmology in a form that may appear primitive on the surface, but truly hid superior knowledge of the working of the firmament of heaven. This enormous knowledge base of the cosmos had the World Tree at its center! In different cultures and belief systems, this World Tree takes the form of the World Pillars. These magical pillars hold up the Earth in a place of the sky. The World Tree is often connected to a magical stone and the sons of thunder, often called the Fire Twins. As we go around the world in search of the revealing sacred stories of the magical Tree, we encounter interesting heroes, fire-breathing dragons, representations of different animals, tricksters, and at least one beautiful princess.

We wonder how the World Tree became a universal symbol around the world, in seemingly unrelated cultures. Was there once a common legend, a postdiluvian mono-myth that everybody rehearsed and carried to all the corners of the world? Is it truly conceivable that there was once a common language, and a common past for all humanity when concerned magi traveled to the far reaches to spread the word and to civilize those who survived the deluge without the required cosmological knowledge? From our history classes, one would assume that the ancients did not know how the globe operated in our cosmos, but the universality of the astronomical tales from different continents that we uncovered suggested the opposite!

Our investigations point to the possibilities that a long-forgotten civilization might have existed and then perished with the sinking of the continent of Atlantis. Certainly, the likelihood exist that previous technologically advanced civilizations—such as Atlantis—attained a much more complex scientific understanding of the workings of the universe than we did so far. Whether it was the relative calmness of the ice age, or that they received information from extra-terrestrial sources is difficult to ascertain. One thing is for sure: whatever evidence of greatness existed was wiped out by the Flood. The coded knowledge that remained with the few survivals is left within the creation stories from all around the world. Can we rediscover that coded system that is based on the commonly agreed shape between friends and foes from Russia to China and from Iraq to the USA?

How do the cosmic world trees fit into this secret model? We think that, by forming a working Cosmic Model, we can rediscover the Lost Secret Formula of our Mother Earth, the sun, and our entire galaxy.

Introduction

In the sacred book of the Torah we hear about the Tree of Life and the Tree of Knowledge:

And the Lord God planted a garden eastward, in Eden; and then He put the man whom He had formed. And out of the ground made the Lord God to grow every tree that is pleasant to the sight, and good for food; the **Tree of Life** *also in the midst of the garden, and the* **Tree of Knowledge** *of good and evil.*

—Torah, Genesis, 1:8-9

We hope to identify these magical trees and discover a unified and easily decipherable chain of astronomical and cosmological events in legends from all over the world about the World Tree. The Tree of Life and The Tree of Knowledge is in the midst of the Garden, thus at least those two are growing on planet Earth, since so far we have not ascertained the locations of other planets with life forms, although we sense that they also exist. One shall then hope to be able to point out the main power brokers of the living Cosmos. We feel assured that, by the use of the Four Directional Cosmic Model, we shall recognize and understand the role of all the cosmic players involved with this magical World Tree and the gathering for the shift of 2012–2013!

We would like to thank all of our families, Patrick Geryl, Andrew Collins, and a number of other scholars, friends, and the editors and book publishers who helped in the making of this book. We would like to extend our greatest appreciations to our wives, Eva Gaspar and Tawnia Jaramillo, who invested so much money, time, and energy for this book to become a reality. Special thanks to Eva Gaspar, who spent valuable time reading the whole manuscript. She and others provided us with valuable insights into the ease or difficulty of understanding our sometimes complex ideas.

CHAPTER 1

The Sacred Feminine VA and the Galactic Heart

> *The universal medicine for the Soul is the Supreme Reason and Absolute Justice; for the mind, mathematical and practical Truth.*
>
> –Albert Pike, Morals and Dogma

The beauty in our human species is that almost every race and culture has a holy book or creation legend they think is so different but, basically, the stories are truly very similar in their core, about creation, eternity, and an almighty spirit god. This is true, from the Siberian shamans to the Sumerian Magyars, all the way to the Babylonian magi and Keltic druids, and from the spiritual Hebrew scholars to the Egyptian and Mayan astronomy priests. For a lot of us, the precious Greco-Roman classical myths can carry an exact account of the ancient changes that happened to the Earth several thousands of years ago. Even the simple and nature-based animalistic rituals of African shamans and Native American medicine men hide well-built and precise astronomical tales that lead us to the UNIVERSAL MONO-MYTH. By actively examining those tales, participating in their rituals, allowed us to solve the cosmic model that opened up a Pandora's Box of ancient secrets that are still pouring our way. We think that by understanding the past and the natural rhythms that the Ancients based their legends and calendars on we can predict the near future.

We believe that the knowledge of Astronomy will provide the importance of the characters of certain animal figures, heroes, and villains and, most importantly, we cherish the Magical Tree standing in the center of our world protecting Creation. Interestingly, only after twelve years of Heroic Journey, meeting the Harlot and then killing the Minotaur and Fire Breathing Dragons can we understand what happened to the Earth almost 12,000 years ago. We need Twelve Tribes of Israel, Twelve Disciples of Christ, Twelve Knights of King Arthur, Twelve Labors of Hercules, the Last Twelve Hours of the Egyptian Book of the Dead and Twelve Years of the Babylonian Gilgamesh to Kill the Bull and Cut out the Cedar of Lebanon. Only by going through all those life-threatening events and then cutting out that magical World Tree, burning everything to ashes and then bringing the flood to wash it off the face of our Earth would allow a new World Tree to sprout and our Hero to Reborn.

Almost every corner of our globe has a tribe or ancient civilization that went to incredible efforts to record their creation stories, almost always including massive earth changes culminating in World Burning and then a Huge Flood. With the approaching date of 2012-2013, we are potentially at the doorstep of similar Earth shaking events.

As our good friend and accomplished author John Jay Harper would state in his excellent book titled *Tranceformers, Shamans of the 21st Century*:

Despite all the marvels of the modern world, these are perilous times. Our world is on the brink of cataclysmic changes that need to be brought to the attention of the general public and media by all of us who have some insight into them.

Our conversation with John Jay Harper goes back over 10 years when we both predicted that soon the issue of global warming will be resolved, since the approaching global cooling and possible axis shift would start a new Ice Age. This was not too long ago, but most scientists would ignore our messages as they expected further warming for decades to come. As our average temperature of the Earth fell back to the 1936 level in 2008, scientists should look at some of our theories that seem to predict the future better than their computer models.

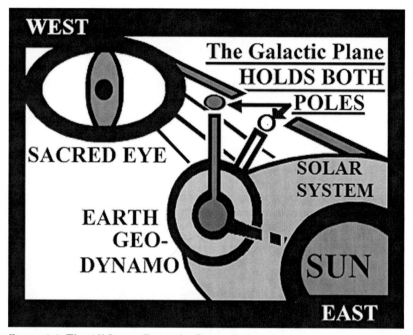

Figure 1-1: The All-Seeing Eye is the Center Bulge of the Super Sun positioned in our Milky Way Galaxy. The Black Hole of our Galaxy is to the west of our Solar System. Modified from a Map of the Milky Way Galaxy from the National Geography

Naturally, we tried to assemble this book recognizing our differences and **celebrating our ONENESS!** We understand that every religion and spirituality carries the seeds of a yearning to understand the wonders of our changing cosmos. We shall demonstrate in our book that the basic elements of the Mono-Myth are found in most creation stories around the world and it means very precise astronomical chain of events and precise timing with predictable earth changes. This World Wide Mythology likely originated from a highly technologically advanced society that existed prior to the Deluge and prior to the sinking of the continent of Atlantis.

We believe that ancient wisdom, sacred teachings, and the precise cosmological messages are not equal to a Rorschach Inkblot Test, where the spiritual teacher can guess about what the meanings of certain animal symbols might be. This will not be the case for our research. Yes, we realize that there are several levels of understanding on the same subjects. There are historical, personal, moral, astrological, and spiritual meanings for those mythological animals. What we are emphasizing here is that, with the current level of scientific understanding in astronomy there is ONLY ONE COSMOLOGICAL MEANING TO THE POLAR BEAR or the EGYPTIAN JACKAL Anubis! The Polar Bear amongst the northerners or the Jackal in the case of the ancient Egyptian society solely stands for the POLAR NORTH! The Swooping HAWK stands for the Magnetic North.

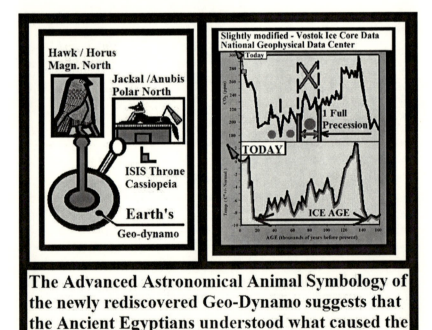

Figure 1-2: The Swooping HAWK or Horus in Egypt stands for the Magnetic North direction of the Core of the Earth, and the JACKAL or Anubis stands for the Polar North. The existence of these two poles of our globe allows the geo-dynamo to produce inner heat for Mother Earth. (Drawing by William Gaspar)

The World Tree 5

The **two animals** that represent the **two pillars** of our Earth are the HAWK and the JACKAL in Egypt. As matter of fact, the main star of the Harp constellation is the Swooping Hawk. The Harp is the only musical instrument amongst the 88 star constellations we find in the sky. According to the science of Astronomy, the star Vega that was called the Swooping Hawk by the Persian astronomers is positioned straight up to the north and represents the Magnetic North, which is also synonymous in Cosmology with the Keystone, the Core of the Earth. To the Polar North sits Polaris, the Polar Bear that was designated to be the Jackal in tropical Egypt. In September in the month of the Virgin we can also detect the Throne of Isis, or Cassiopeia near the Polar North. Anybody who has seen the Egyptian Horus recognizes the fact that the Hawk Headed Prince is often sitting on the Throne and wears the Sun with a Serpent on its head. These important relationships of cosmology is what we will crack open.

In most legends there are Two Brothers or Two Sisters, the Old King, the Queen, the Sacred Baby of the Sun, the Virgin and the Harlot, the magical fire-breathing Dragon and the evil Serpent. The Bull Minotaur plays an important role in the Mithraic Mysteries of the pre-Christian Mediterranean's and to this day the Bull fight remained an ancient ritual of great importance for the Latin cultures of Spain and Mexico. In the Nativity scenes of the Christians the Cow, Donkey and the Dog have a constant presence around the sacred baby. The Bee and the Beehive, the Cat, Lion, or Jaguar is extremely important. The Dog, Fox, Wolf, Goose or Wild Turkey is also major players in the theater of world mythology. These animals and others, along with the heroes, heroines and villains seem to be part of most legends and will help us decipher the astronomical secrets of the Ages.

This is what we learn from the world-renowned Catholic scholar in his classic book, *Patterns in Comparative Religion* about the World Tree of Nordic mythology:

The prophetess, the VÖLVA awoken from a deep sleep by Odin to reveal the beginning and end of the world to the gods, declares: I remember giants born at the dawn of time and those who first gave birth to me. I know nine worlds, nine spheres covered by the tree

of the world. That tree set up in wisdom which grows down to the Bosom of the earth. I know there is an ash tree they call Yggdrasil.
—Mircea Eliade, *Patterns in Comparative Religion*

First, we are taken aback by the bluntness and the clearly vulgar appearing name of this Scandinavian prophetess, the **VÖLVA!** We just cannot imagine that holy men and holy women throughout our long history would have allowed a derogatory title for the seer of the Earth changes that reduces her to one important body part. This spiritually and mythologically lasting name for the Vagina of this Nordic Prophetess has to contain so much Wisdom of Cosmology to survive the tests of time that we decided to spend more time on exploring her secret. We assume that the language of Sex and Vulgarity of the Wise Ancients were cleverly employed to discourage average seekers to touch the subject in its entirety. Therefore, the tales could stay open to the public

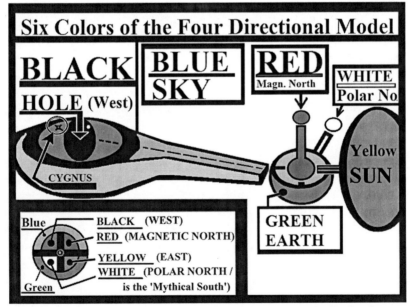

Figure 1-3: The FOUR DIRECTIONAL MODEL Here to the 'Del' side view shown on the horizontal plane of the Solar System. The two main views will be the above and the Secret one from the west originating in the heart of the Super Massive Black Hole.

The World Tree 7

and kept an important secret in plain sight. The clichés of sexually oriented secrets, such as the 'Pussy Cat' and the 'Diamonds are girl's best friend' will be explained in later chapters.

Once the legends entertain half animal half god creatures who mate with humans or even just Harlots with lose morals who fornicate with the kings of the Earth what insane seeker would try to obtain a sacred understanding of God's creation in that tale? Certainly, we the authors would like to do that. The perceived **vulgar** tone of the sacred tales do not alienate us rather, we start suspecting that these legends must hide enormous cosmic enigmas that can only be discovered if we apply our best minds to research most creation stories and hidden metaphysical knowledge to **find the meaning of this important female body part as a cosmic principle (VA)**.

We based the Four Directional Model on the ancient colors of the Lakota Indians, the Egyptians, and the Mayans. Although, out of the Four Main Colors of the Galaxy the Two Colors of Mother Earth and her Two Pillars, the RED and WHITE is present on most flags of various countries around the world. Even Santa Claus is in favor of only those two colors. This might explain why the Old Man of Winter set up residence at the North Pole – one of the Two Pillars of Mother Earth.

The Two famous Pillars that hold up Earth in the big empty sky are the Magnetic North (LaVA-red) and Polar North (Snow-white). Just observing the Four Directional Model of the Earth's geo-dynamo with its two North Poles, we sense that we shall be able to solve how the Sacred Feminine "VA" from the 'völ-VA' developed and will fit into our theory.

The ancient astronomy priests who originally developed this symbol were in possession of the real Cosmic Model and its alignments, and certainly understood human psyche extremely well. They picked a vehicle (VA) that seemed sacred and sturdy, and also one that is surrounded by enough desire to build and destroy countries, decide kingships, and motivate people—from the poorest to the most noble—to be able to give everything up and to follow the call of nature.

In the ancient cosmic language of signs and symbols, there existed a complex cosmologically-based vocabulary in which the VOL or VUL syllables denoted a Cosmic Female Birth Canal that was conceptually connected to the perfect VA-lley and the VOL-canism from VUL-can and a few other concepts, such as the Wolverines and the Scandinavian Volvo / 'turning' concept. Certainly, the Vulcan name was the title of one of the Greek gods and when those unruly deities began their wars against the unsuspecting heroes of humanity then huge earth changes happened.

From the earlier quote about our Nordic prophetess, Völva, one can certainly sense that the story is about the end of a cycle and the beginning of another. In later chapters, discovering an ancient system of solving the enigma of cosmic sexual union, Völva will give up her secret in a very specific, albeit symbolic, fashion.

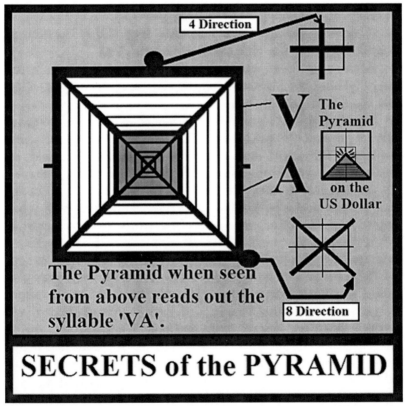

Figure 1-4: The bird view of the Pyramid displays the letter 'X' that also reads out the 'VA' when the X is sliced in half on the Horizontal plane. We noticed that it also provided the letter A as the enigmatic Pyramid with the cut off top we find on the US One Dollar bill (Drawing by Wm. Gaspar)}

The World Tree 9

This VA is the Sacred Valley of cosmology that shall be more than just a place to us. She will be much more than the biblical harlot who fornicates with the kings, as is implied in the book of Revelation of the Bible. She will be the sacred feminine of this beautiful globe we live on, and her painful fertility role will slowly roll out of our worldwide legends. Could she be Mother Earth or her mythical sister, the planet, Venus? Maybe she is the First Woman of Native creation lore in the heart of our galaxy. Why not think of her as the first woman born to the Judeo-Christian tales whose name is the Hebrew E-VA.

We began to recognize that a number of legendary female and deity names end in VA, such as E-VA (Hebrew), Ší-VA (Hindu), Völ-VA (Scandinavian) and Miner-VA (Roman) and we felt that we were on the right path. We also know from ancient Magyar tongue that there is a spell-bind connection between the Valley, Lady and the Oak Tree (Völgy/Hölgy/Tölgy) as all three words spelled very closely suggesting a linguistic cosmic relatedness. Thus, the Magyar VÖL-gy (=VA-lley) and the Scandinavian Viking word of VÖL-VA have a close relatedness then will explain why the VA-gina and the VA-lley connected in our upside down woman called Mother Earth,

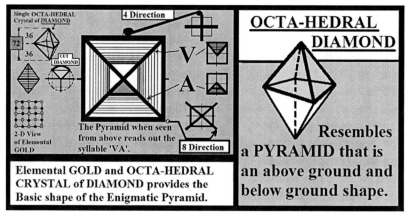

Figure 1-5: *The shape of the PYRAMID buildings that we so readily find in Egypt and Mexico was derived from the knowledge of the Elemental structure of GOLD and DIAMOND before the Big Flood and the sinking of the continent of Atlantis. That in itself is a proof to us that ancient civilizations were highly developed technologically. (Drawing by Wm. Gaspar*

Thus, we begin our search for the 'VA' syllable in other Creation legends and feminine related concepts to arrive to the conclusion that we will find much more words than just **VÖL-VA, E-VA, O-VA, Di-VA, Ši-VA and Miner-VA that relate to this Cosmic secret.** Naturally, even these early discoveries provided us with an important reassurance that our search is progressing in the right direction!

It cannot be a mere coincidence that the elemental structure of Gold and Diamond was used to represent Mother Earth in the secret ancient formula of the Pyramid. The eight sided double pyramid represents the sacred number 8 and also reminds us of the concept of the 'as above so below'. The top pyramid represents Mother Earth on the surface and then the mirror image of that pyramid below the surface likely stands for the Core of the Earth. This suggests that the pre-diluvian Atlanteans had to possess a technologically superb society before the sinking of their continent almost 12,000 years ago. They had to be far more advanced if they readily understood and employed the knowledge of the elemental shapes of metals and precious stones along with the workings of the Earth's geo-dynamo.

Their science had to be superior to ours due to the fact that they were able to erect huge stone pyramids in such accuracy that most builders today would not be able to reproduce. Certainly, the glass and metal objects broke since and rusted away, the fine electronic equipments disintegrated, but the enormous sized blocks that they used to build some of those structures with, are still standing there tall in honor of the skills of the ancient masons. Those huge blocks were not pulled by half naked slaves. The likes of levitation, magnetic motors, plasma, and hydrogen engines had to be in the possessions of the Atlanteans to do the work that needed to be done. At least they poured the fine cement in place without dragging blocks. The history that we have learnt in our class rooms do not explain the complexity of our past prior to the Deluge.

Certainly, the Secret Formula of the Sacred Feminine of Mother Earth was assembled with such a superb cosmic knowledge, simplicity and universality amongst the Ancients that it points to a civilization that passed us in knowledge and spirituality. They did not ascend out of a dark cave as historians would have us believe. The caveman stage of human existence might have arrived for us after the Flood

The World Tree 11

wiped out almost everybody and everything. Before that Flood, a golden age of humanity must have existed that was capable of long distance interplanetary travels and other magical things we cannot conceive. To assemble and maintain a complex secret cosmic language and letter making formula that in one form or another was preserved in all corners of the world definitely speaks for past greatness.

We certainly gained scholarly amazement about the VA as our First Sacred Feminine emerged out the shape of the pyramid that was based on elemental gold and diamond. She is a virgin-like being as the Egyptian Isis Mother, but she also has the qualities of a sensual being Hathor, the Harlot, who behaved with a sense of looseness. Just judging from the ancient tales, she must be old, very old! In the Magyar tongue, we call very old VÉN, a word that is almost pronounced as the English VAIN and maybe a reminder for the planet VENus. In Spanish and English, VEN-ison refers to the meat of the DEER. So we are NOT surprised that by the conclusion of our book this cosmic hunt will end up in the ORION ZONE where the daughter of Orion (Orion=Hindu Prajāpati), the DEER called DAWN (conceived by incest) will be born. That is the specific beginning in the earth change scenarios. We are planning to analyze the world mythologies and calendars until we know exactly when and how to expect the approaching cataclysm.

Continuing on with a few more special words let me mention that the Spanish / Latin / Magyar VÉNA word, that means the blood carrying VEIN in English. This VÉNA also linguistically connects us to the idea of the VEN-ous blood that flows during the time of cosmic menstruation. But this is not a flesh-and-blood, human menstruation; rather, this is the menstruation of Mother Earth that is related to ORIGINAL SIN. This "blood" connects us to the menstruation that happened, according to the Hindu myths, while Mother Earth was making "cosmic love" to the incarnate WILD BOAR aspects of Vishnu, who is known in Hindu myths as the Pervader, originally a solar god, but then becomes a supreme ruler.

First of all, there is a god / goddess named Vena in the Rig Veda of Hindu Myths whose identification with the Sun, Soma, high ocean waves and a celestial birth of a calf.

Next, we observe in the *Hindu Myths* by Penguin Classics the idea of a menstruating Mother Earth who was full of desire for the fiery seed of the Lord of the Universe:

> *The earth, in the water, is full of desire and behaving like a woman; she has been violated and has conceived a cruel embryo from your fiery seed; she is menstruating, however, and therefore unfit for the embryo with which you impregnated her, O lord of the universe.*
> —Hindu Myths, Penguin Classic

A Cosmic Sex scenario is opening up to us where the female partner is Mother Earth and the male participant is the Lord of the Universe himself! So, at least now we know that the VA (VÖL-VA) is a part of Mother Earth who is making cosmic love to the LORD OF THE UNIVERSE!

Let us continue and find out if a baby was born out of this cosmic union? A few more sentences from this tale of the Káliká Purána:

> *Therefore the son who will be born will also carry off unrestrained women; he will develop a demonic nature and harm the gods and Gandharvas [celestial musicians, authors]. Thus the lord of the world spoke to Vishnu, who had a shameful look, about a vicious son begotten by sexual pleasure with a menstruating woman, a son who would do unpleasant things: "Abandon this lustful boar-body, lord of the world; you alone are the cause of creation, preservation, and destruction; you are the cause of the world . . ."*
>
> *But the boar went to his own mountain, named World-non-World, and he made love with the earth, who had the form of a lovely female boar, and though he made love with her for a very long time on that supreme mountain the lord of the world was still unsatisfied, as his lust for the female boar was so strong.*
> —Hindu Myths, Penguin Classic, pg. 188-189

We learn from the above quote that a 'vicious son' had been born from the Cosmic Union! We would like to know his birthday so we can prepare for his return:

In the Saiva version of this Hindu myth, this is how Ši-VA

The World Tree

chastises the boar:
> Even in the Vaisnava story of the boar avatar there are hints of incidental destruction caused by the boar... and explains the boar's fatal flaw in social, ritual, and cosmic terms: he becomes too attached to his family, makes love to a polluted woman (as Agni / "the Fire god" makes love to the Krittikás /Pleiades when they are monstrous, according to some versions), and is unable to control his destructive fire once it has been loosed. On the first level, the earth herself shares his ambivalence: both of them know what should be done, but they are unable to do it without the help of Šiva. On the ritual level, the polluting "blood" is expiated by a blood sacrifice of the boar . . . Doomsday is imminent:
>
> Vishnu takes the form of a fish and saves the Vedas, as he does at the final flood."
>
> —Hindu Myths, Penguin Classics

From the above quote, one need to remember that this event is put together with the Hindu Fire God, Agni who makes cosmic love at the Krittikás (Pleiades) where the Boar is on the astronomical chart. The Seven Sisters of Pleiades is the Shoulder Blade of the Taurus Bull in the sky, thus we need to pay special attention to that since the Toreador or the Matador, who is ritually risking his life fighting the Bull in the ancient Latino cultures is sacred. When we checked the date of the Pleiades, the Seven Sister – the Killing of the Celestial Bull on the Egyptian calendar of the Zodiac of Dendera we realized that it fell on January 1st ! Then, we began to suspect that the specific date of the burning of the Earth was tied to the 'birth of the Sacred Child of the Sun'. This Son of the Sun in the Hindu myths would be found in the tales of Krishna or Lord Skanda.

> The supreme purification of this entire universe is to be accomplished by ashes; I place my seed in ashes and sprinkle creatures with it. ... By means of ashes, my seed, one is released from all sins. ... I am Agni of great energy, and this woman, my wife Ambiká, is Soma; and I am both Agni and Soma, as I am myself the

Man together with Nature.
–Penguin Classics, Hindu Myths

Certainly, not desiring to be ambiguous about the source of the Fire, we search the Hindu Myths to find out from where this 'vicious son' would be arriving?

When the noble sages had performed the sacrifice in the proper way, they offered the oblation into the Oblation-devourer kindled with good fuel for all the dwellers in heaven. The marvelous fire, the carrier of oblations, the lord, was summoned there; he came out from the orb of the sun, ...
–Penguin Classics, Hindu Myths

Now, we are very certain about the fact that the world mythologies secret language will be talking about a Hero, THE LORD, as it is stated above, CAME OUT OF THE ORB OF THE SUN!

Naturally, it makes good sense to have the Son of the Sun, as this is what the Egyptians keep calling the Hero, come from the Sun, as that obvious cosmic scenario explains why everybody would be able to see that 'illuminated Son of the Sun'.

Then as we feverishly searched for more clues about the 'Fire' in mythology we found out that the Fire started as "angry bees and flying feathered serpents" arrived from the sun in the Fourth hour of the last twelve hours of the Egyptian *Book of the Dead*. Studying the depictions it was even more surprising that those half-naked Egyptians carried a sacrificial Boar in their legend of the Judgment related to this changing period! Thus, in this multi-cultural mythology search we already detected close relations between what the Mayans, Egyptians and Hindu myths would store for us.

Now, after that faithful year in the past when the Sun erupted and burned the Earth to ashes in the fourth years of the last twelve of world mythology, we noticed that it was first followed by abundant water and then the infamous *Lake of Fire* in the Fifth Year. Not only this Lake of Fire is presented in the Bible, but it seems to involve the magic of turning water to wine or even into blood. Well, blood is an

The World Tree

important media in most religions and spiritualities and it is part of the command and the sacred Covenant with the Lord in both the Torah and the New Testament. Even the frightening book of Revelation places the 'lake of fire' as the second death. That is, because the fourth hour's Sun eruption causes the first big mass death and then the fifth hour's axis shift causes the second death.

14 Then Death and Hades were cast into the lake of fire. This is the second death.
–Bible, Revelation 20:14

The Lake of Fire was the time when after the Magnetic Axis shift caused the underwater volcanoes to erupt and their rusty sulfur content colored the water. Thus, when the Pharaoh did not allow the Jews to leave Egypt, Moses threw his rod into the river that turned red. Cosmologically, that Rod was the Iron Rod of the Magnetic North. The blood related ritual became ritualistic even amongst the ancient Jews.

6 Moses took one part of the blood and put it in basins, and the other part of the blood he dashed against the altar. 7 Then he took the record of the covenant and read it aloud to the people. And they said, "All that the Lord has spoken we will faithfully do!" 8 Moses took the blood and dashed it on the people and said, "This is the blood of the covenant that the Lord now makes with you concerning all these commands."
–Berlin & Brettler, *The Jewish Study Bible*, Tanakh Translation

Thus, the magical covenant of Adon, the Jewish Lord was made with the people of Israel. The amazing history and the resilience of the Hebrew tribes from the extended areas of Egypt remains one of the marvels of ancient and even current written recorded history.

17 'Thus says the LORD: "By this you shall know that I am the LORD. Behold, I will strike the waters which are in the river with the rod that is in my hand, and they shall be turned to blood.
–Bible, Exodus 7:17

The critical last 12 years of the Cataclysm that culminated in the Great Flood really perked up our interest. Is there a way to date the exact year when the last cataclysm happened and when the new one could? How did the Ancients tried to convey that to us? Where do we start? Maybe understanding how the cycles are broken down in the Ice Ages is one thing, but what about first cluing in on the Iron Rod that God gave to His favorite kings to rule over the land. Is it a magic wand of our Creator or is simply an enigmatic language of the old bards to describe the way the Earth's geo-dynamo works?

Now, this famous 'rod' of Astronomy and Cosmology shares a possession between Mother Earth and the Sun King. We only recalled one important metal component of the Earth, which was its Iron Core, thus the Rod that emerged from the Core of the Earth was the Magnetic North. So far it seemed fitting and we also found another VA or WA in the name of the Egyptian Iron Rod that was titled 'WA-S' or 'VA-S'. The symbolic representation of this Iron Rod is also found on the Christmas tree in the form of the Candy Cain that is commonly decorated with the colors of red and white.

If the ancient songs are about our Mother Earth and her demise, then everything that we will learn about mythology shall be translated into the workings of the inner heat producing Geo-dynamo of the Globe and how it is controlled from the west and the east. This 'Iron Rod' that is emerging from the Iron Core of the Earth and divides the Galactic space in half between the westerly Northern Cross and the easterly Sun is a major pillar of our cosmic scenario. We will spend time on this concept since the VA-S iron rod is supported by the Sun, and its fall decides our faith periodically. When God / Allah or the Great Spirit decides to 'throw' that Rod to the ground then the axis shift began and the waters of the oceans shall take on the color of blood or wine.

This Iron Rod that is called VAS / WAS in the Egyptian hieroglyphic language is extremely important to our theory, not only because it fits our VA linguistic test, but because it is a Staff that was depicted as the medical Caduceus Staff with wings. Since the Swooping HAWK perches on top of that Magnetic North / Iron Rod we sense that the wings were that of the Hawk. We realized at that point that we have even seen angels of mythology depicted with a harp and with hawk

The World Tree 17

wings on a painting somewhere. Where was it? We forgot! We can't find anything when we need it the most.

Well, after an insane search through a number of Art History books, works of famous painters, we finally stumbled on the right one. It was in the book edited by Deborah Cannarella titled Christmas Treasures. It was truly a treasure. It was a painting from the Pinacoteca, Vatican Museums – another secret VA – where Melozza da Forli painted the Music-making Angel. One could call it the Cosmic Angel of the Magnetic North! This Angel was a Red-headed Lady (the Red color represents the Magnetic North) who was playing on the Harp

Figure 1-6: The Iron Rod of Mythology in the Bible is the Magnetic North Pillar emerging from the Iron Core of the Earth. In this view as we are observing the Earth we sense that this Iron Rod rules our faith and our lives (Drawing by Wm. Gaspar

(the Harp also represents the Magnetic North) and she wore on her back, Hawk wings (the Swooping Hawk star Vega also stands for the Magnetic North) - but on top of all this revelation about the Magnetic North, this red-headed angel had the yellow halo of the Sun behind her head to tie this whole Iron Rod business to the Sun.

This famous painting was housed in the Vatican Museum where the Cosmic Angel displayed at least three major markers of secret cosmology standing for the Magnetic North / Iron Rod / Keystone Core of the Earth and all of it tied to the Sun. What was the most amazing thing about this is that the famous painter Melozza da Forli lived and painted during the centuries of the Dark Ages. Naturally, a flat Earth does not need a Magnetic North or a Core and certainly not a dynamo. Just as a tire, if the dynamo is flat it does not spin – and we know that our Earth spins.

This dichotomy of cosmic knowledge hidden in religious art remains a great enigma not only for the Dark Ages, but for all religions and spiritualities of all times.

It started feeling like someone is hiding something and we did not understand who and why. We instinctively knew to begin to look for answers where few would search for the advanced knowledge of Astronomy and Cosmology. We began searching for light in the gloomiest appearing times of recent history - we looked for answers in the religious art of the Dark Ages.

The Dark Ages probably began when the Roman Latin Empire - that already inherited the cosmic teachings of the Persians, Babylonians, Jews and the ancient Egyptians – now set its sight on Jerusalem that was a Holy City of the Jews, Muslims and the Christians. When the First Crusaders captured Jerusalem and established a Latin Kingdom with their first king as the French Godfrey of Bouillon, the power struggle became more intense between Christians and Muslims. When the advancing Turks generally exterminated the combined German and French forces of the Second Crusade with Louis VII, the French king barely escaping, it should had been a sign for the Christians to forever avoid fighting for that region. The popularity of the Knight Templars of the Crusades fluctuated with their successes and failures.

When we dig deeper into the Knight Templars' history, it seems that it was not the cosmic knowledge or the roundness of the Earth

The World Tree

that came between the Pope and his secret society rather it was the greed of the French King Phillip the Fair. The story of the Knight Templars or otherwise known as The Order of the Poor Knights of Christ and the Temple of Solomon began in 1128 when Hugues de Payens founded the organization to guard the road to Jerusalem for the pilgrims of the Catholic faith. The Holy See of the Vatican sanctioned The Poor Fellow Soldiers of Jesus Christ to secure the road to Jerusalem that was slowly taken over by the army of Saladin, the great general of the Saracens. For a while, these not so poor knights were able to hold off the advances of the Muslim horsemen and were able to maintain their Headquarters in the city of Jerusalem, in the Dome of the Rock that was built by Muslims.

The Knight Templars, who were at the forefront of the Crusades, began to lose their popularity as they were losing the war for Jerusalem. This was the time when the 23rd Grand Master of the Knight Templars, Jacques De Molay came to power. When Pope Boniface, VIII declared that the French king had no right to tax the Knight Templars who officially were the servants of the Papacy - that created a strain between the French King Phillip IV and the Italian Pope Boniface. Certainly, a lot of rich nobles of France took advantage of this tax credit and joined the Knight Templars, further robbing the French king of cash reserves. After, Guillaume de Nogaret, the willing henchman of King Phillip kidnapped an important Bishop – the furious Pope Boniface VIII excommunicated both Phillip IV and his loyal soldier. This religious and financial power struggle erupted into a vicious war between the French King and the Italian Pope. Shortly after that event Pope Boniface was kidnapped and held by the French King for 3 days. During that time of the imprisonment, the Pope was apparently severely beaten and died a month later. His unlucky successor, Pope Benedict XI also died of suspected poisoning by de Nogaret only after a year as the Pope at the Vatican. The next Pope, the Frenchman Clement V, who was handpicked and groomed by the French king Phillip IV, in an unexpected action moved to Avignon, France thereby throwing the Vatican into chaos.

Finally this era opened up the possibilities for Phillip IV to deal with the rebellious Knight Templars, who until then only answered to the Pope and no earthly kings. After seven years of torture and the brutal execution of the charismatic Grand Master of the Knight

Templars Jacques DeMolay on March 18th 1,314 AD on the Island of the Jews - the Cosmic Knowledge went underground. The hidden secrets of Astronomy and Cosmology that was always the domain of the Priesthood in past great empires now became disconnected of the Papacy. Thus, during the worst power outages of the Dark Ages the secret societies were forced to establish prominent art schools throughout Europe to keep the cosmic knowledge alive.

One of the main cosmic secrets was the action of the Iron Rod of Mother Earth that now with the alienation of the Order of the Knight Templars eluded the Vatican. The stories of the Torah and the Bible were kept alive, but the secret knowledge of its astronomical meanings officially was separated from the Church. It was now the responsibility of the Secret Societies with or without the help of the Pope to keep the cosmic knowledge alive. The Bible still talked about the Iron Rod and the role of Venus in the calendar counting, except the religious majority had and still has no idea of what was being said.

The Rod of Iron of the Magnetic North and the Morning Star

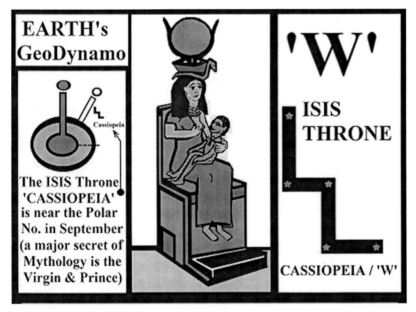

Figure 1-7: Cassiopeia in the sky is the Throne of Isis, the Queen Bee representing the famous celestial letter W of astronomy. (Drawing by Wm. Gaspar

The World Tree

of Venus still seemed important in the Bible, except nobody could translate it anymore.

27 "He shall rule them with a rod of iron; ... "
28 "And I will give him the morning star"
-Bible, Revelation 3:27-28

Well, with important historical understanding we began our search for the Cosmic Mother, for the planet Venus and a Queen in the sky that could further our limited understanding of the workings of the Milky Way Galaxy.

Searching for at least One Astronomical Queen in the Egyptian Sky, we remembered the magical Throne of Isis shone as the Cassiopeia star constellation in the night sky. It is seen near the Pole Star, Polaris in September in the month of the Virgin when one looks north. As we will later realize, the Throne of Isis has a very important mythological function, as it is the place where Horus, the hawk-headed young prince, the SON OF THE SUN would be crowned as the new SUN KING. We understood that there had to be a mythological and cosmic interrelatedness of the Egyptian Queen to the Judeo-Christian tales and to the Cosmic Mother Earth. The Sacred Cosmic Marriage of the Hindu Wild Boar to Mother Earth began to yield fruit.

Let us be glad and rejoice and give Him glory, for the marriage of the Lamb has come, and His wife has made herself ready.
—Bible, Revelation, 19:7

We stopped and rested. After a little siesta, we decided to halt searching for more females for now and we gladly returned to complete the discussion on the significance of numbers.

"The number Eight is the Number of Years it takes the planet Venus to go around the sun. Remember, it takes eight years for the VENUS TRANSITS of the "two sisters" to occur around the sun. The first Venus Transit of our century happened on June 8, 2004, and then the second one will be Eight Years later, on June 6, 2012. That is two days short of eight years! That is why we think the eight is a significant number, since the Mayans fervently monitored and recorded down the cycles of our sister planet, Venus.

We remembered that the planet Venus took Eight years and Five stations to go around the sun once, from our earthly perspective. Both the five and the eight are numbers in the Fibonacci number sequence (0,1,1,2,3,5,8,13, ...) of spiral growth and likely important counting

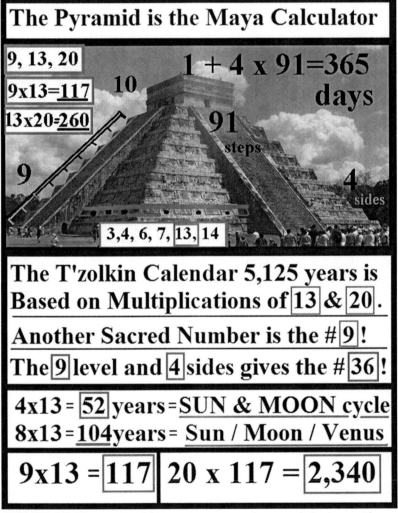

FIGURE 1-8: *The ancient Mayan astronomy priests used the pyramid as a calculator! (Photo Credit Eva Gaspar)*

The World Tree   23

measures. Noticing the planet Venus taking eight years around the sun we realized that we needed to compute with the number eight! We also remembered that in the Buddhist spiritual believe it was The Eight Fold Path that a seeker needed to follow.

We were certain that these numbers existed in various denominations around the world for maintaining a reminder to a cosmic knowledge of natural rhythms.

8 x 9 = 72: A FULL ONE DEGREE of PRECESSION!

While the FIXED STARS OF THE SKY turned ONE DEGREE in the firmament of the heavens in 72 years, the planet Venus circled around the sun NINE times. Therefore, we realized that both the 8 and the 9 are Earth-Venus-Sun numbers relating to the One Degree of Precession! It almost seems that the number 9 is related to the cycles of the Sun, the planet Venus is definitely owns the number 5 and 8 and Mother Earth has to be known by the number 7, as it takes 52 x 7 days to go around the Sun once.

"We can see those numbers. Did we pass your kindergarten requirements?" Tawnia and Eva giggled listening in on our conversation.

We looked at each other, shaking our heads. "No! There is much more!"

"Even for a kindergarten grade?" Tawnia and Eva laughed as they ran after Tawnia's little girl, Maya.

"Yes, a whole bunch more." we assured them.

Pepe turned back toward me. "I am not going to talk about the Venus numbers of 117 and 2340 in the first chapter of our book!" Dr. Jose Jaramillo stated with conviction as we reviewed the higher multiplications of these natural numbers.

"I agree. Let's first build up the framework of the Mayan and the Egyptian astronomy and cosmology as we see it in comparison to the Ice Age rhythms.

There were many more calculations than we show here, but for our purposes, that Mayan priests used the number 52 YEARS as a marker of time. Out of the four brightest lights in the ancient sky, the two most luminous, the sun and the moon, met at the point of the Pleiades every 52 years. That was the year of the HARMONIC CONVERGENCE popularized by José Argüelles. The last year that 52-year meeting took place was in 1987. Even today we calculate 52 weeks to be One Year around our Sun, thus we tie the One Year Cycle of the Sun to the number 52!

Thus, with the unusual spelling and presentations for the meaning of the Sun the Egyptian RÁ Priests hoped that we will understand the

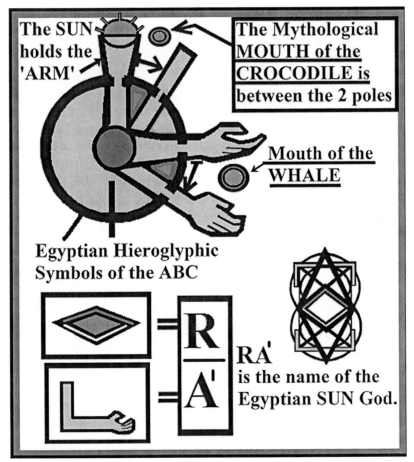

Figure 1-9: The Egyptian spelling of the Sun consists of the letter 'R' (='Mouth') and the letter 'A' (='Arm'). Our book demonstrates clearly why the ancients Egyptians did not simply pick a Round Circle with Rays coming out of it (Drawing by Wm. Gaspar)

The World Tree 25

complex wisdom they hid behind the simple body parts as letters.

The Sun, the Moon and the planet Venus was important marker in the sky together. Therefore, if we involve the third brightest light in the sky, the planet VENUS, those THREE bright ones—The Sun, the Moon and planet Venus—meet every 104 YEARS!

The fourth brightest light in the ancient sky was the Dog Star, Sirius – the 'Scorcher' as the ancient Greeks named it - and this Sothic Sirius Cycle is about 1,460–1,461 year long. These numbers, along with the 117, 2340, 96, 100, and more will become important later!

We sat down to reflect on the Force that turns our Earth from the west to the east. This region is symbolically identified by the Goose or the Northern Cross in the HEART of our Super Sun of the Milky

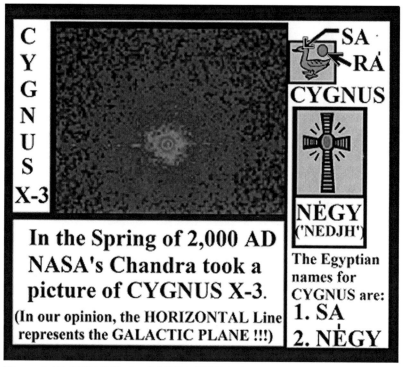

Figure 1-10: CYGNUS, the GOOSE, SWAN, shown here as the RESONANT CROSS, seems to be the proposed HEART and POWERHOUSE of our Milky Way Galaxy! (Drawing by Wm. Gaspar)}

Way Galaxy. It is in the active, embryonic region of the Super Massive Black Hole where most ancient sacred births originate from. We began to compare the actual rhythms in the Ice Ages to the numbers used by the ancient Mayan astronomer priests. The Mayan numbers kept showing a detailed knowledge of these Natural Cosmic Rhythms that are scientifically recorded down in the Ice Core samples of Antarctica (Vostok) and Greenland (GRIP / GISP)!

As NASA showed us in the spring of the year 2000, there is an area around the Super Massive Black Hole of our Milky Way Galaxy that we mythologically identify with the Heart of the Goose that is the star Cygnus X-3 or X-1.

Only since about 1972, when the scientists began to study the Gamma Ray radiation emanating from Cygnus X-3 that is the star in the Crossing of the Northern Cross. This assumed binary system, with Cygnus X-3 in its center, from our point of view is thought to be not only a very powerful micro quasar, but likely the strongest gamma ray emitter source that turns our Earth from the west toward the east. The fact that the gamma ray radiation has a 4.8 hour periodicity, in our opinion makes Cygnus X-3 even more likely to be the originator of the force near the Galactic Plane that in five successive waves daily (5 x 4.8 = 24 hours) turns our Mother Earth toward the Sun. Recently, the Italian astronomer Marco Tavani's research provided valuable research into the understanding of this important micro quasar.

Naturally, if Cygnus was not an important constellation in the west that has to do with the alignment of the Sun and the subsequent eruption and burning of the Earth then the Classical Greek Mythology would not connect the two celestial bodies, the Cygnus Swan or Goose and the Sun, but they did. According to one Greek legend Cygnus, the Swan was best friend with Phaeton the son of Helios, the SUN GOD. One day PHAETON, the Son of the Sun asked his famous father to let him drive the Chariot of the Sun. Helios was reluctant first, but then gave in to the pleading cries of his son. Naturally, Phaeton's reckless driving of the Chariot too close to the Sun threatened the Earth to be burn to ashes by the heat of the sun. Zeus, the main god intervened and shot a thunderbolt at Phaeton who fell into the River Eridanus (starts at the star Rigel, the foot of Orion) that is still known as the 'steaming river' due to Phaeton's fall. Phaeton's best friend, Cygnus

The World Tree 27

dived into the river to save Phaeton. For his brave action Zeus made Cygnus into a Swan and placed him in the sky.

Now, this Greek myth clearly connects the alignment of the westerly Cygnus and the easterly Sun and also warns us about the overheating of Earth from the Sun. Furthermore, because all of this happened at and around the constellation Orion, the tale brings in the River Eridanus that begins at Rigel, the foot of Orion. Even more, the newer poetic tales of Smith refers to the impending Axis shift of the Magnetic North to the Polar North in a southwardly fashion by revoking the alternative title, the Northern Cross name of the Cygnus constellation and connecting it to the name of the Sacred Baby of the Sun - who was actually cosmologically born from the alignment of the Swan, the Northern Cross with the Sun. If Smith, the writer was a pious uninitiated Christian, the Northern Cross of Calvary designation would be sufficient without the mentioning of the Swan and its flight to the south. Thus, Smith was likely a good Christian with some knowledge of the Cosmic Mysteries Christ would only teach to His disciples:

Yonder goes Cygnus, the Swan, flying southward,-
Sign of the Cross and of Christ

Now, if someone thinks that these connotations are made up and Cygnus has no mythological connections to the Magnetic North (LYRE / Harp / Hawk / Stone) then we would like to offer the following quote from the beautiful book of Skywatching by David H. Levy:

One story claims that Cygnus is Orpheus, the great hero of Thrace, who sang and played his lyre so beautifully that wild animals and even the trees would come to hear him. It is said that Orpheus was transported to the sky as a swan, so that he could be near his cherished lyre.

Thus, the celestial music of the earth changes began. It should be progressively more evident, as we review the world wide mythology that the same cosmic elements prop up in every religion and spirituality, thus we begin to sense that interconnectedness of ancient cultures that were teaching us the same basic history, morality and cosmology.

One last thing we would like to demonstrate in this first chapter of our book is the fact that this Northern Cross of world mythology existed in Egypt, in the Mayan creation legend much before the advent of Christianity. Therefore, we assume that the advanced scientific knowledge about the importance of the star constellation of Cygnus, the Swan, the Galactic Goose or Northern Cross has existed for several thousands of years. This cosmic realization of how the Geo-dynamo is controlled by the Cross from the west, and the Sun from the east might have been around for over 50,000 years or longer. The accumulated knowledge of cosmology about the recurrent earth changes that caused the repeated near annihilation of humanity over and over again - created a scientific and spiritual store house of languages, religious signs and symbols that we employ in our everyday lives most of the time without the actual recognition of what they stand for in the sky.

The colors of the Olympic Rings also display the five of the six

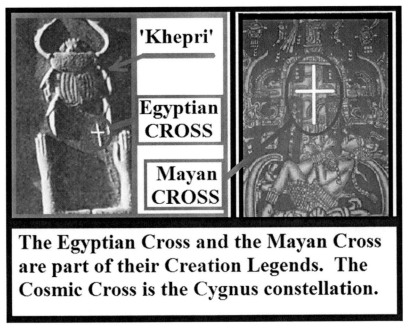

The Egyptian Cross and the Mayan Cross are part of their Creation Legends. The Cosmic Cross is the Cygnus constellation.

Figure 1-11: The examples of the Galactic Cross at the center of the Creation legends were present in ancient Egypt and with the astronomy minded Mayans much before the advent of Christianity. (Drawing and photo by Wm. Gaspar)

The World Tree

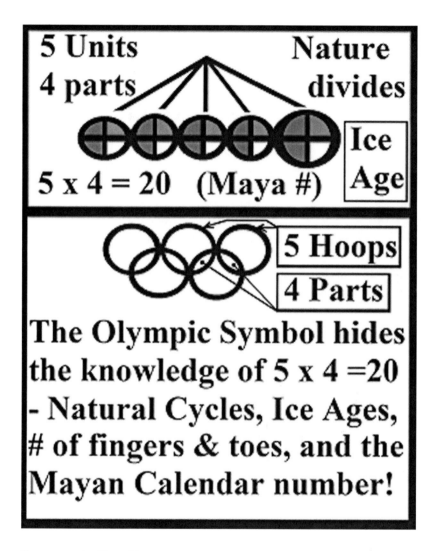

Figure 1-12: The Olympic Symbol of World Unity the Five Circles interlocking at Four Meeting points strangely represents the Five Precessions of a ~120,000 year Ice Age with Four Divisions within each Precession. This multiplies out to the sum of 20, reflecting both the number of reversals in an Ice Age and the sacred Mayan number found in the T'zolkin Calendar (Drawing by Wm. Gaspar)

colors of The Four Directional model. The rings are Blue, Black and Red on top and Yellow and Green on the bottom. The only color missing is White that they are using for the back ground. The Blue color naturally stands for the sky and the Green is the color of Mother Earth in full bounty.

We need to begin to comprehend that most celestial symbols, colors, signs and languages of Cosmology is part of our shared past as the descendants of the surviving members of the multitudes of earth changes. These ageless symbols are not there to divide and separate us from the One Uniting Power in the Universe that is an Almighty Spirit Force or Creator that gave us lives. Maybe at the end times of each major earth changes the gates become narrow and limiting, thus those of us who will be able to crowd through it together will have to learn to respect, love and help each other. We shall walk hand in hand as proud reminders of outstanding specimens of an extremely rare and special species called the Human Race.

CHAPTER 2

The Rock, the Red Hero and the Precession of the Ice Ages

> *The disciples of Hermes, before promising their adept the elixir of long life ... advised them to seek for the Philosopher's Stone ... This Stone, says the Masters in Alchemy, is the true Salt of the philosophers, which enters as one third into the composition of Azoth.*
> —Albert Pike, Morals and Dogma

If one remembers that in this chapter the only thing that is important that the Force from the West ('SA') holds the Pillars of the Earth ('K') and the Sun from the East ('RA') stabilizes or destabilizes the Rock Core inside our Mother Earth, then we achieved our goal.

The ŠA-K-RA, meaning 'Powerful' in Hindu mythology or the SA-C-RA-ment of the Judeo-Christian teachings, both refers to this powerful west-earth-east alignment holding the Rock inside the globe stable. Every 12,000 years the West to East alignment through the heart Rock of Mother Earth is rolled out of its place and causes the major earth changes. This Sacred Rock is embedded deep in the heart of Mother Earth and we try to visualize that an Iron Rod is emerging on top of this Iron Core of Mother Earth and it shoots straight up to the Magnetic North direction and the top end of this Pole sits a Hawk that is the deciding factor in the severe earth changes.

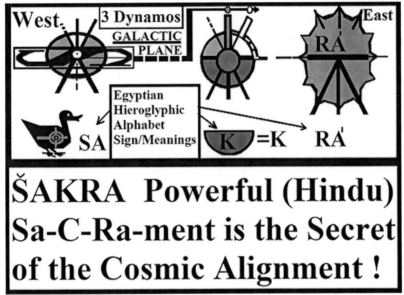

Figure 2-1: SA-K-RA is the Sacred Alignment from the West (SA) through the Rock & Two Pillars of the Earth (K) all the way to the Sun (RA) (Drawing by Wm. Gaspar)

When we learn the parts of the Cosmic Mysteries then it will be easier to assemble them later on. It is refreshing to see that the logo of the upcoming 2010 Canadian Winter Olympics is the Inukshuk that is a stone human figure, a worthy representation of Hercules as the Giant who will move the Stone inside Mother Earth. Even more revealing is the sacred Totem Pole of the Native American Indians from the northwestern region of North America and recall that most of the time the top of that carefully carved totem pole ends in a Hawk or an Eagle with open wings ready to take a flight. When the flight of that Swooping Hawk happens then the Totem Pole falls. That is our rudimentary introduction and explanation for a much more complex science.

How amazing it is to witness the presence of the exact parts of Cosmology from the American Indians to the Egyptians and from the Hebrews to Christians and the Nordics. The Book of Job naturally mentions the Swooping Hawk, but even in other books of the Torah, the Tanakh and the Bible we will find evidence of this cosmic

The World Tree 33

knowledge of the west and the east destabilizing the Core of the Earth. The knowledge of the presence of the Magnetic North and its periodic instability will not always be represented by the Harp, Hawk, Iron Rod or the Rock inside Mother Earth. Sometimes, it will be symbolized from above as a standard or a plumb hanging down from a rope of the north and provide us with a sense of instability by a pendulum like motion. In the quote below from the book of Isaiah, one can detect the same advanced science that quite innocently weaves the knowledge of the power of the west and the easterly Sun tied to a standard that is usually a small banner on a high pole. Another common way to look at the Magnetic North is as a plumb iron weight hanging down from the North by a Plumb Line.

> *19 So shall they fear the name of the LORD from the west, and His glory from the rising of the sun; when the enemy comes in like a flood, the Spirit of the LORD will lift up a standard against him.*
> —Bible, Isaiah 59:19

Rather than reading the Old and the New Testament of the Bible as a translation about an angry old God, we realized that we have advanced star knowledge of astronomy from the ancient Hebrews and the Latinos. Thus, truly the tug of war in between the west and the east outside the globe is what decides the fate of Mother Earth. The westerly resonation from the Black Hole holds both North Pillars and the Sun stabilize the Magnetic North Pillar, the Iron Rod. When searching for Cosmology in the Four Directional Cosmic Model inside creation legends and holy books, the information is honestly this simple.

> *THEN Moses answered and said, "But suppose they will not believe me or listen to my voice; suppose they say, 'The Lord has not appeared to you,'"*
> *2 So the Lord said to him, "What is that in your hand?" He said, "A rod."*
> *3 And He said, "Cast it on the ground." So he cast it on the ground, and it became a serpent;*
> —Bible, Exodus 4:1-3,

Naturally, one or two quotes will not define the cosmic mysteries of the Torah, the Tanakh and the Bible or any other holy book. It is the gestalt of the world mythology that will prove our point right. One of the best books to read on the Egyptian Gods and compare them to our biblical wisdom is the one written by E. A. Wallis Budge.

XVI. 1. *Thus Heru-Behutet and Horus, the son of Isis, slaughtered that evil Enemy, and his fiends, and the inert foes, and came forth with them to the water on the west side of the district. And Heru-Behutet was in the form of a man of mighty strength, and he had the face of a hawk, and his head was crowned with the White Crown. ('Polar North' – auth.) and the Red Crown (Magnetic North – auth.) .2. And they slew the enemies all together on the west of Per-Rehu, on the edge of the stream (Rigel –auth.) ... Now these things took place on the 7th day of the first month of the season PERT. ... and the Lake which is close by it hath been called TEMT ('Temet'-auth.) from that day to this, ... called the FESTIVAL OF SAILING ...*

Then Set took upon himself the form of a hissing serpent. Let "Horus, the son of Isis, in the form of a hawk-headed staff", set himself over the place where he is, so that "the serpent may never more appear."

–E.A. Wallis Budge, *Legends of the Egyptian Gods, The Legends of Horus of Behutet*

There is a lot of valuable information in the above quote. First of all, the Hawk and the Hawk-headed staff is the Iron Rod of the Magnetic North, represented by the Swooping Hawk. The unification of the White Crown and the Red Crown is the unification of the Polar North (Snow –white) and the Magnetic North (Lava-red). This has to be in the FIFTH or the 'Lake of Fire' Hour of the Last Twelve Hours before the Flood. Thus, the Unification tells us the year when the Axis Shift happened. The other clue to that is the LAKE word that refers to the Lake of Fire of the Egyptians and the Bible. The Temet (='Sled') word again covers the Fifth Hour when this event start sliding down symbolizing the axis shift and the severity of the earth changes. As matter of fact, those who would like to visually double check that fifth

The World Tree

hour with the Sled, could see the pictorial representation the best in Erik Hornung's book titled The Ancient Egyptian Books of the Afterlife. The obvious double oxymoron of the situation is the similarity of the 'temet' Sled representing the destruction and resembling the Sled of Santa Claus cosmologically, and the third fact is that historically there was no snow in tropical Egypt.

Still analyzing the above quote, in the last sentence the 'serpent' and the 'staff' is connected to the Axis shift. It happened when Horus, the Swooping Hawk flew south to sit on the Throne of Isis.

Therefore, the 'hellish' Lake of Fire is connected to the earth changes as much as the next quote is tying Satan to the Rod of Iron and the dashed pieces of the potter's vase, which represents the burnt and broken up Mother Earth. As matter of fact, the potter's vessel or the Vase is another VA of cosmological significance.

24 "Now to you I say, and to the rest in Thyatira, as many as do not have this doctrine, who have not known the depths of Satan ...

25 "But hold fast what you have till I come.

26 "And he who overcomes, and keeps My works until the end, to him I will give power over the nations-

27 'He shall rule them with a rod of iron; they shall be dashed to pieces like the potter's vessels' ...
–Bible, Revelation 2:24-27,

Thus, during an alignment when the Sun erupts, following that the ROCK inside Mother Earth loses its stabilization and ROLLS toward the South. Observing this from the westerly Black hole – we would see the Iron Rod of Magnetic North shift from Noon to the One O'clock hourly position on the face of the Celestial Clock! This chief 'rock & roll' cornerstone is the Rock Core / Iron Rod of Mother Earth.

"6 Therefore it is also contained in the Scripture, "Behold, I lay in Zion a chief cornerstone ...

7 "The stone which the builders rejected has become the chief cornerstone,"

8 And "A stone of stumbling and a rock of offense."
–Bible, I Peter 2:6-8,

Any holy book is usually written first with the historical aspects of a war or struggle and the moral laws emphasized, to be aware of the past and to master for now how to live an honorable life. The deeper secrets of an exact Cosmology are always there as seemingly meaningless words to the uninitiated, but can be very revealing to those who search for the astronomy and cosmology amongst the lines.

In astronomy the star constellation Hercules is just below the Hawk and displays the 'cubical' shaped Keystone body that represents this mythical chief cornerstone, the Rock. The secret representations of this Rock are preserved for the worshippers in numerous legends and locations throughout our history. The cubical shaped Rock that holds King Arthur's Excalibur is not any different than the holy Ka'aba (='Cube') stone visited in Mecca during the Hajj by the Muslims.

Another famous holy rock is kept in the Dome of the Rock on the Temple Mount in Jerusalem. Since, Abraham was the father of both the Jews and the Arabs we can assume that the name of the stone kept in Mecca, the Arabic Ka'aba (Cube) and the Jewish Gabbatha (Pavement) where Jesus was judged came from the same ancient Semitic root word referring to the cubical Keystone. This cubical Keystone thus connects us Jews, Christians and Muslims and further on the rest of the world.

"13 When Pilate therefore heard that saying, he brought Jesus out and sat down in the judgment seat in a place that is called The Pavement, but in Hebrew, Gabbatha."
–Bible, John 19:13,

We found hundreds of examples for the 'stone' word in the Bible and we assumed that the secret cosmic meaning was referring to the knowledge of the 'Keystone' that was inside Mother Earth. When someone had been 'stoned' as a punishment in the Bible, we do not necessarily have to imagine an angry lynching mob, instead we can lovingly think of the biblical fathers who assembled the Scripture as caring elders wanting to warn us about a larger stone inside the Earth that will hurt us. A lot of the tales are not about angry killers, but about sacred teachers willing to seem childish and ridiculous just to be able to forewarn us with their stories. Once we are able to detect the sacredness of these legends, then in place of the ethnic hatred that

The World Tree 37

was based on false assumptions, we could develop a warm and fuzzy feeling about the ancient Jews and the early Christians who sat down alongside with other shamans to create a Holy Book that spoke to all of us and try to warn all of us. As we search for more wisdom from the Bible, we found another very specific quote referring to the hot lava aspect of the Magnetic North in the book of Jeremiah. This hot boiling pot of the second death of the fifth hour is coming out of the Magnetic North.

13 And the word of the LORD came to me the second time, saying, "What do you see?" And I said, "I see a boiling pot, and it is facing away from the north."
14 Then the LORD said to me: "Out of the north calamity shall break forth on all the inhabitants of the land.
—Bible, book of Jeremiah 1:13-14,

Naturally, a number of people raised on the conventional wisdom of Sunday school teachings of the Bible would not think of the Tree, as the masculine power and the stone inside Mother Earth as the Feminine birth giving force. Luckily, the Torah and its inherited wisdom in the Old Testament of the Bible gave us a few pointers.

24 A wild donkey used to the wilderness, that sniffs at the wind in her desire: in her time of mating, who can turn her away? ...
27 Saying to a tree, 'You are my father, and to a stone, 'You gave birth to me',
—Bible, book of Jeremiah, 2:24-27,

Besides finding the references to the Sacred Tree as the Male Power, it is refreshing to see that the turning action of the Stone inside Mother Earth is identified as a birth giving parent. Besides unwillingly learning about the specifics of the mating style of female Wild Donkeys, this biblical quote provided us with a well needed clue to the timing of the Birth of the Son of the Sun. All we had to do is, to find a Wild Donkey full of sexual desire that was tied to the birth along the Ecliptic Path of the Sun. That was easy, as we recalled a pair of Donkeys positioned in the body of the star constellation Cancer. Thus, we recorded in our memory that a Donkey and a Sacred Birth was tied together in the sky. We wanted to be ascertained about the timing of the beginning of the earth changes, along with the hiatus on the Path of the Sun and the obvious burning of the Earth directly from heaven.

14 And a messenger came to Job and said, "The oxen were plowing and the donkeys feeding besides them ...
16 ... "The fire of God fell from heaven and burned up the sheep and the servants,
–Bible, book of Job 1:14-16,

First, we did not want to believe that God is known by sending fire, but on further readings we were convinced that He does.

24 "Then you call on the name of your gods, and I will call on the name of the LORD; and the God who answers by fire. He is God."
–Bible, book of Kings I, 18-24,

Therefore, we accepted that the birth of the Son of the Sun is as fiery as the anger of our Lord, who sends fire. We had now a list of things we could bring together in the name of the Lord. The birth pangs beginning at the time of the Bull and the actual birth of the Son of the Sun when the Donkey stars appeared on the Ecliptic Path of the Sun gave us some good pointers as far as the ancient disaster was concerned.

We needed to tie the turning of the magical Stone to the birth of the Lord and the time of the astronomical Donkey. That did not seem like an easy task, but we wanted to know more. We were searching for examples of the Hero and the Stone.

"Now on the first day of the week Mary Magdalene went to the tomb early, while it was still dark, and saw that the stone had been taken away from the tomb"
–Bible, John 20:1,

The World Tree 39

Was this sacred stone the same rock that the ancients Greeks tied a Virgin to off the beach, when the Beast from the Sea insisted on the human virgin sacrifice given to him? Was it the brave Heracles, the famous hero Hercules who saved the daughter of the King from a certain death? Is it the same allegory that is used in all other mythology? We certainly believe so.

One of the most sacred Rocks or Stones for both the Judeo-Christians and the Muslims is inside the Dome of the Rock in Jerusalem, Israel. For the Muslims, a much more important Rock is the Ka'aba that is found in Mecca. The Rock of Jerusalem also carried a monumental history for the believers in the Prophet of Islam. We will write some more about those later, but right now we would like to bring attention to two words from the possible secret cosmic word mysteries of Christianity. It is certainly not an oxymoron how it was intended initially, but the interpretations today lack the cosmology. The two words from the above quote by John are the **Tomb** and the **Stone** that is at the entrance of the tomb where Jesus was laid to rest after the Crucifixion. The Tomb is usually a common mythological tool of the Ancients to define the Dark Rift in the Galactic Bulge. This Dark Rift was named Xibalba Be (Black Road of the West) by the Mayans and it is mostly where their Creation Legend is played out by the Two Brothers.

Well, this Dark Rift houses the Northern Cross – the Goose of Astronomy, the westerly power that is able to turn the Stone (Core) of the Earth. We can certainly find the Cave or Tomb allegory with the travels and adventures of the Greek Flood hero Ulysses, or otherwise known as Odysseus, who lands on the island of Sicily with twelve companions in search of food. Passing by huge flocks of sheep, they arrive to a large Cave that is filled with rich stores of cheese and milk (Milky Way reference). This mysterious cave is guarded by the son of Neptune (Ocean reference), Polyphemus who was the meanest of the One-eyed Cyclopeans (Galactic Center reference). The One-eyed clue is about the Galactic Center power that can be seen on our first picture of this book. The Cave or Tomb reference is about the Dark Rift where the Goose – Northern Cross is located. Since, the Dark Rift contains the Goose - Cross power of the west and the Sun from the east, thus it is the only force that can roll the Stone of Mother Earth.

We understand that a number of religious people do not think that there are Cosmic Mysteries hidden amongst Christ words, but for those who are in doubt – we need to say that we believe that secret meanings are hidden in between the lines and when even Jesus Christ attests to that then one needs to believe.

9 Then His disciples asked Him, saying, "What does this parable mean?"
10 And He said, "To you it has been given to know the mysteries of the kingdom of God, but to the rest it is given in parables, that 'Seeing they may not see, and hearing they may not understand.'
–Bible, Luke 8:9-10

Thus, there is a great possibility that in the tales of all the ancient heroes and redeemers, we encounter the same hidden tale of the Savior Heroes who teach us about important cosmology through their own legendary sacrifices that most people will not clearly understand as far as the hidden astronomy is concerned.

For God will bring every work into judgment, including every secret thing, whether good or evil.
--ible, Ecclesiastes12:14

Since we are examining the mysteries of the Cosmos, we are not worried that the secrets will be misleading. Our goal is to find the knowledge of the natural rhythms of the Universe hidden in science and amongst the pages of the holy books and creation legends.

To begin with our search, we needed to find a scientific proof that the rhythms of the Universe were indeed connected to the Force from the west. We noticed that there was an interesting correlation between a human heart beat and the Ice Age cycle graph. As matter of fact, they looked almost identical. Could a human heart beat be originated from the Super Massive Black Hole of our Milky Way Galaxy? If the Galactic Goose-Cross is what turns the Earth from the west to the east and define the Sun cycles, then we could conclude that the Ice Age cycles are the results of the Spiral Force of the Galactic Heart. Thus, we are not surprised to observe that the Heart Beat of Mother Earth - that is

The World Tree 41

~117,000 years long Ice Age period – is practically identical to the 1 second long human Heart Beat.

Since, the Heartbeat of an Ice Age Cycle of 117,000 years and the

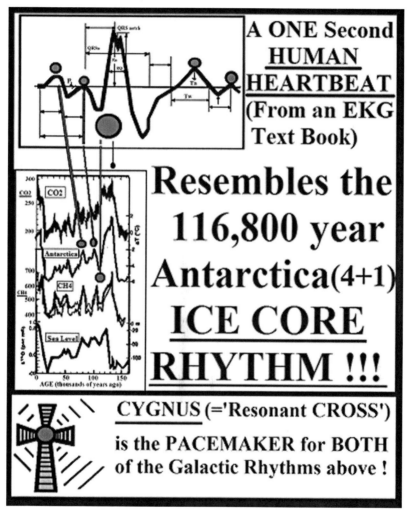

Figure 2-2: The one second long human heartbeat and the heartbeat of Mother Earth is practically identical. One can find the partitions of 5, 10, and 20 in those graphs. The Precession of the Equinoxes carves out the path of the Ice Ages. (Graph by Wm. Gaspar)

one second long human heartbeat are essentially morphologically identical; then we derive the origins of all rhythms from the heart of the Milky Way Galaxy. The excellent book by Andrew Collins titled The Cygnus Mystery, subtitled 'Unlocking the Ancient Secret of Life's Origins in the Cosmos' came to mind. In The Cygnus Mystery Collins discussed his thoughts about the Mayan Long Count Calendar, the World Tree, world mythology and an ancient cosmic knowledge:

The precise relationship between the end of a former age, or 'sun', in Mayan cosmology and the beginning of the current one was reckoned by a complex calendrical system known as the 'Long Count'. ... All that can be added with any certainty is that – like the Phoenix in Graeco-Egyptian legend that returns to Heliopolis and is consumed in its own pyre at the end of each world age – Itzam-ye / Seven Macaw came to signify the culmination of one sun, and the commencement of another. As such, it is my belief that cosmological myths of this nature, involving a celestial bird, the World Tree, and the use of stars and constellations to monitor cosmic time, have a universal origin that reaches back to Paleolithic times."

We not only whole-heartedly agree with Collins, but are searching the world mythology to submit further proofs to that effect.

Andrew Collins further stated that:

It is the opinion of a number of well-informed writers, including John Major Jenkins, author of the highly recommended Cosmogenesis 2012, that the upcoming galactic alignment is something that the Maya not only planned for 2,300 years ago, but also described in the Popol Vuh story of the Hero Twins heralding the birth of the new sun. Jenkins' understanding of Maya cosmology is second to none and he builds on this order to strengthen his case. For him, the Milky Way's Great Rift, the 'Road to Xibalba', can be seen as a kind of 'cosmic birth canal', out of which the all-important sun will emerge on 21 December 2012.

Certainly, we understand that the stars that make up the Goose star constellation are not actually a unit of stars that sit around the Black Hole, but by conventional wisdom, the Ancients agreed that the shape of those stars that we see from here represents the imaginary "heart" of our Black Hole Power. The fact, that the Ancients widely knew this important scientific information thousands of years ago, is another proof that the Atlanteans had to possess great technological knowledge. Our scientists only began to understand some of this science beginning with the space age few decades ago. Only in the last 15 years or so is when NASA publicly acknowledged that we have a great westerly gamma ray source from the area of Cygnus X-3 and X-

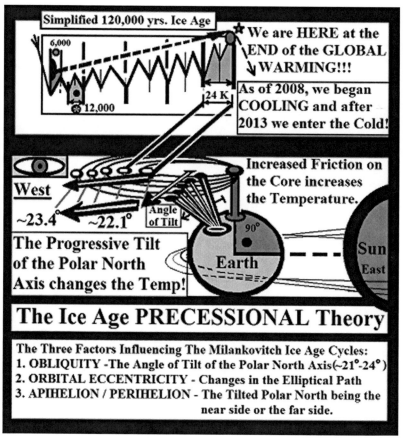

Figure 2-3: *The gradual and progressive tilt of the Polar Axis of Earth is the Precession. This Precession is what determines the specific cycles of the Ice Ages (Graph by Wm. Gaspar)*

1. How is it possible that the Ancients knew this to a higher degree widely distributed before the Great Flood?

The Cygnus Goose (SA) from the west and the Sun (RA) from the east control the Rock of the Earth, the Magnetic North axis and thereby defining the length of the Great Year of Plato or what Hipparchus in 127 B.C. called the **Precession of the Equinoxes.**

This is what Giorgio de Santillana & Hertha Von Dechend wrote about the ambiguity of the Precession and the discovery of it by Hipparchus in their now classic work titled *Hamlet's Mill*:

There is good reason to assume that he actually rediscovered this, that it had been known some thousands of years previously, and that on it the Archaic Age based its long-range computation of time. Modern archeological scholars have been singularly obtuse about the idea because they have cultivated a pristine ignorance of the Precession itself.

Then de Santillana and Von Dechend continue in Hamlet's Mill state the following:

We today are aware of the Precession as the gentle tilting of our globe, an irrelevant at that. . . Today, the Precession is a well-established fact. The space-time continuum does not affect it. It is by now a boring complication. It has lost relevance for our affairs, whereas once it was the only majestic secular motion that our ancestors could keep in mind when they looked for a great cycle which could affect humanity as a whole. ...

But then our ancestors were astronomers and astrologers. They believed that the sliding of the sun along the equinoctial point affected the frame of the cosmos and determined a succession of world-ages under different zodiacal signs.

Thus, we suggest that the 'sliding of the sun along the equinoctal' and solsticial points every about 6 -12,000 years - as the alignment of the Sun with the heart of the Black Hole causes the axis to shift and the geo-dynamo to be recharged. The official theory and measurements

The World Tree

of scientists estimate a 10,000 to 20,000-year time period for the Magnetic Field of the Earth to dissipate. The recharge of the Magnetic Field at those times happens fast and involves an Axis Shift. In the past the Magnetic North Axis shifted to the Polar North Axis. In the position of the Polar North they united and later separated. This event is the basis of the countless Creation Legends and Ceremonies about the Sacred Cosmic Marriage ceremony of the Ancients and the going into captivity and returning from it.

This Winter Solstice Alignment of the sun with the Galactic Heart is where the Mayan calendar ending of 2012 comes into focus. This is based on the Galactic Alignment theory popularized by our favorite Mayan scholar, John Major Jenkins who was more accepting of our Ice Age Theory being connected to the Super Massive Black Hole than the scientists of the 1990s. His now classic work, The *Maya Cosmogenesis 2012*, detailed the **Galactic Alignment of the Winter Solstice Sun with the HEART of our Galaxy.** This is how *John Major Jenkins* writes about the Croll-Milankovitch Ice Age Theory in his next book the *Galactic Alignment*:

> *I also want to mention in this context the reevaluation of the Milankovitch theory of ice ages. The idea that the advance and retreat of polar ice is related to precession was put forth over a hundred years ago by the Scottish thinker James Croll. Milankovitch reanimated the ice age-precession theory in the 1920s, but only recently has science reconfirmed Croll's convictions with the overwhelming new evidence, vindicating his work. Now, taking this all a step further by embracing the galactic alignment phenomenon, Dr. William Gaspar has rekindled and expanded the implications of the Croll-Milankovitch model in his recent book, The Celestial Clock.*

Thus, if the **Black Hole-driven Ice Age Theory** is accepted, then we are talking about a new understanding of the Theory of the Ice Ages. Combining the Mayan calendar 2012 ending with the ice age cycles was a revolutionary idea and it gained the attention of others.

We remembered that the MAYAN T'zolkin Long Count calendar is 5,125 years long. Since the number 8 is assigned to the planet Venus and the Mayan calendar was largely based on the measurements of the planet Venus, then we have another revelation about the correlations between Natural Rhythms in the Ice Age cycle and the Mayan calendar:

That is: **8 x 5,125 years = 41,000 years.** The 41,000-year cycle (18,000 + 23,000 = 41,000) is a very important period in the

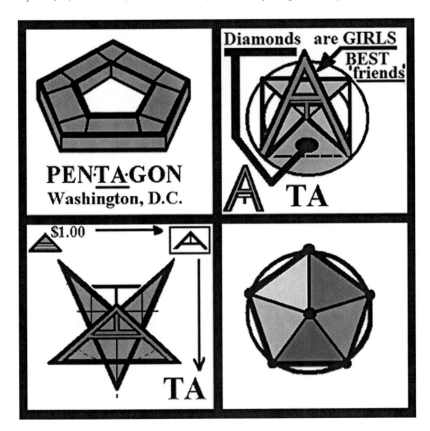

Figure 2-4: The Pentagon shape and the Pentagram are present in cosmology and astronomy. PEN-TA is one of the Secrets. "Pen" has masculine and punishment connotations and the "TA" ending equals to the "VA" in importance as the Sacred Feminine of the Cosmic Union. That is why a male saint is only 'San' such as San Diego and the female saint is 'San-TA' Maria! (Drawing by Wm. Gaspar)

The World Tree 47

Figure 2-5: The sacred TA is the secret representation of the Sacred Feminine, the Cosmic Union and the Birth of the Son of the Sun. 'TA' is also the biliteral word assigned to the 'Belt of Orion' sign in the Egyptian alphabet.

The Painting is by Dieric Bouts titled The Adoration of The Angels. It is located at the Museo del Prado, Madrid, Spain. (Photo credit: Erich Lessing / Art Resource, NY)

Milankovitch Ice Age cycles. Furthermore, it also suggests to us that the advanced scientific knowledge of Cosmology may have been around for a long time. The Galaxy and the Cosmos does not change in its basic Spiral Pair character, thus the knowledge that was known let us say 50,000 years ago is still valid today. The fact that we find the sum of the 41,000 year long Milankovitch cycle amongst the Mayan numbers is a reassurance that civilization is much older than we initially thought.

Even Geoff Stray, the now world renowned author of the book, *Beyond 2012: Catastrophe or Ecstasy: A Complete Guide to End-of-Time Predictions,* talks about the Milankovitch Ice Age Cycles in Chapter 9 of his book and he spent the entire Chapter 10 on Dr. Gaspar's theory called Ice Cycles. More and more authors are recognizing the correlations of nature's natural cycles found in the ice age cycles and the rhythms recorded in the very accurate Mayan calendar.

Another surprising sacred numbers that the ancients hid for us in the natural cycles of the Half Precessional Ice Age is the 11,680 years. **When the Half Cycle Precession of 11,680 years divided by the Sacred 20, the number of toes and fingers and also the number of the Mayan calendar, they yield the Magic number of the planet VENUS: 584!**

11,680 / 20 = 584

We understand that the earth changes do not exactly have to happen at 11,680 years since the rumbling last for decades. What we cherished in the divisions of the 11,680 year long Half-Cycle Precession was the fact that it seemed to record down the multiplied sum of 5 (Spiral Creation) and 4 (Directions) = 20 parts of the Half Cycle = 584 years. The 584 year mirrored the 584 days of Venus that built the Pentagon shaped 'house' around the Sun in eight years. Even the advanced defense forces of the United States chose the Pentagon Building to be known by.

One of the main secrets of world mythology and religious lore is the KEYSTONE, and the TA. The TA is the roof above San-TA Maria. Besides the VA, now we can add the TA as one of the secret representations of the Sacred Cosmic Feminine. The TA also hides the secret knowledge of the cut off top of the pyramid that can be readily found on the US One Dollar. In the Egyptian hieroglyphic language SA

The World Tree

is the west power, TA is the Axis Shift that began in Orion, and the N is the Resonation that comes out of the easterly Sun causing the shift. It spells SA-TA-N!

"14 And no wonder! For Satan himself transforms himself into an angel of light."

–Bible, II Corinthians 11:14

Before we become too familiar with the Devil, we shall leave the TA for later and for now we will keep our conversation with the mythological Keystone. Not only the disciples of Hermes, but every major religion and spirituality held the 'sacred stone' in high regards. The stories are not always easily revealing in an exact fashion of astronomical science, but the basic building blocks of the parts of the universal Mono-myth are always there. It may be the Blacksmith and the Mill of the Finnish Kalevala, or maybe the old King who loses his kingdom to the young Prince, or Horus with a Sun on its head. Nonetheless, the story is commonly about the Geo-dynamo of the Earth that is undergoing its timely reversal at the half point of the Precessional cycle. The stone and the pillar concept are necessary to understand for the geo-dynamo of the Earth.

14 So Jacob set up a pillar in the place where He talked with him, a pillar of stone;

–Bible, Genesis 35: 14

This Magical Pillar of Stone cosmologically is hidden in the Bosom of Mother Earth. This Keystone of mythology is the Core of the Earth what makes it possible for our globe to produce volcanism and inner heat, thus in essence allow meaningful human life on Earth to exist. This Keystone is the Cubical shaped Body of Hercules in Astronomy shown near the Magnetic North. This astronomical Hercules is personified in countless stories of mythology and religious tales. It is difficult to present a Galactic Alignment with the razor-sharp gamma ray radiation causing the axis shift. It occurs naturally, in a sense it is a need to reenergize the weakening geo-dynamo of the Earth. Without that switch, life would stop existing on Earth due to the unchecked radiation effect. Thus, a bunch of flesh beings have to disintegrate prematurely every 12,000 years for a few thousands humans and animals to restart and propagate for the rest of the next 12,000 years.

This repeats itself over and over again as part of the cycles of dying and rebirth. The Horizon Project DVD by Brent Miller was partly based on Dr. Gaspar's theory and it is a great visual tool to see the Solar System swinging in and out of the Crucial Alignment.

Therefore, other than the Hawk-headed Horus, the only significant famous Hero of Astronomy is Hercules who helps the deciphering of the cosmic struggles. The Hercules star constellation is near the Magnetic North, thereby representing the Keystone of the Core of the Earth.

Is there a way to prove scientifically that all heroes of creation legends, ancient mythology and religious lore are tied to the concept of the astronomical hero Hercules? Naturally, there is no unequivocal proof, only mountains of implied evidence pointing in that direction. Our quest shall continue to uncover the layers of the universal mono-myth buried under tons of betrayal, treachery, indecent sex, moral preaching and bloody wars between well meaning human heroes and heroines against the Titans representing the unbridled forces of Nature.

We shall continue unabated our attempt to place the most important animal figures of Astronomy around the Earth – those that are most closely related to the Axis Shift – into their proper places.

The 'Two Edged Sword' in the hand of the Cat is the Scissors of the Cosmic Crab. This is about the Magnetic Axis shift thousands of years ago. Now, we can study the astronomy of the CANCER (CRAB) star constellations for clues. 'SARTAN' is the Hebrew name for the CRAB. 'Sartan' is one letter more than 'Satan' or 'Santa' and implies a linguistic aspect of cosmology. When we remember that the westerly force, the Egyptian Goose is named 'SA' and the easterly resonation from the Sun represented by the Egyptian letter 'N' and between the two rivers of resonation is the Belt of Orion called 'TA' - that will clearly represent the birth of the Son of the Sun and the ensuing axis shift on Earth, and with this foundation under our belt we could gradually begin to conceptualize the important letter combinations of the Cosmic Language.

The CANCER star constellation was called the GATE OF MEN by the Platonic philosophers and the Chaldean magi. In this position of the sky the Ancients imagined that the souls arrived from Heaven to occupy the human bodies. Certainly, with all the earth changes happening there at specific times, the soul traffic had to go also to the other direction and

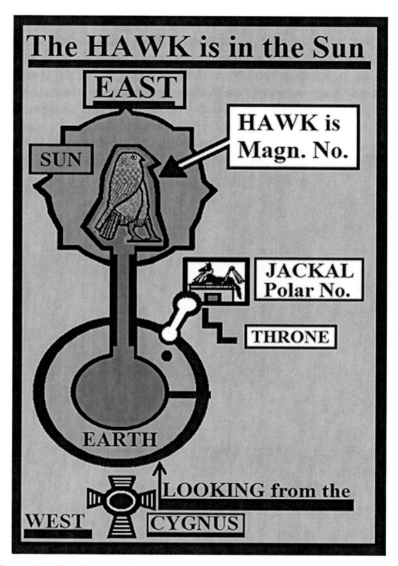

Figure 2-6: The Swooping HAWK of Astronomy that is the Egyptian HORUS shown sitting here on the PILLAR of the Magnetic North that is above the Magical Stone. From this most Sacred direction of the Birth Place of stars we begin to understand visually why the Hawk and the Sun is connected. (Drawing by Wm. Gaspar)

that carnage on the flesh level was worthy of recording.

According to the excellent book titled Skywatching by David H. Levy - 'In Greek mythology, Cancer was sent to distract Hercules when he was fighting with the monster Hydra. The crab was crushed by Hercules's foot, but as a reward for its efforts Hera placed it among the stars'.

So, the Crab was the scissor in the paws of Bastet, the Egyptian Feline Goddess, the Lynx Cat of Astronomy who had to use the two edged instrument to cut off the head of the dangerous Water

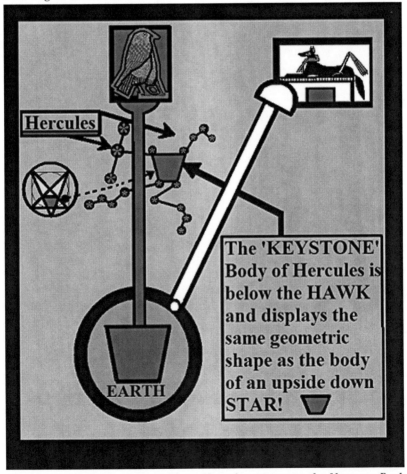

Figure 2-7: The Astronomical Hero, Hercules represents the Keystone Rock of the Core of Mother Earth. (Drawing by Wm. Gaspar)

The World Tree 53

Snake Hydra. Was the water snake Hydra the enigmatic mythological representation of the Huge Tsunami that decimated the shores thousands of years ago? Is the 'RA' syllable in the Latin and English names of these mythological animals, such as Hyd-RA, Cob-RA, Vipe-RA, RA-ven, C-RA-b, D-RA-co, G-RA-il and even the Ly-RA - connects them to the Egyptian Sun God RÁ?

Thus, the Egyptian Cat Bastet stood for the Lynx constellation above the Cancer. We searched for wisdom about the Egyptian Cat to see if it had anything to do with the Magnetic Axis shift, or the so called Unification of the Two Egyptian Kingdom.

9 *The chapter of casting a spell on the Cat*
"Hail Rá, come to thy daughter! A scorpion hath stung her on a

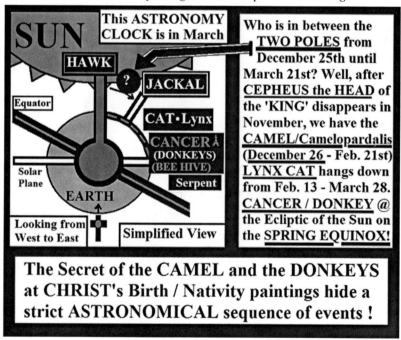

Figure 2-8: The astronomical animals of the Axis shift as they line up on the outline of Mother Earth observed from the Secret View of the West identified by the Goose / Northern Cross at the edge of the Dark Rift of the Black Hole. (Drawing by Wm. Gaspar)

lonely road. ...

16 Thy fear is in all lands, O Lord of the living, Lord of eternity. O thou Cat, thy two eyes are the Eye of the Lord of the Khut uraeus, who illumineth 17 the Two Lands with his Eye, and who illumineth the face on the path of darkness. O thou Cat, thy nose is the nose of 18 Thoth, the Twice Great, Lord of Khemenu (Hermopolis), the Chief of the Two Lands of Rá,

—E.A. Wallis Budge, *Legends of the Egyptian Gods,*
Legend of the Death of Horus,

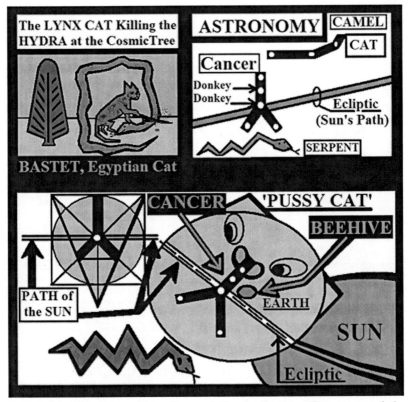

Figure 2-9: The Cat goddess, Bastet, kills the Serpent. This is one of the greatest secrets we solved with our Four Dimensional Cosmic Model. See the Astronomy of the Cat, Cancer and the Hydra Serpent. (Drawing by Wm. Gaspar)

The World Tree

Now, the Cat in Egypt is tied to the Unification of the Two Lands that is the reference to the Magnetic North Axis shift to the Polar North, due to the power of the erupting Sun.

At the end of the quote we find Rá, the Sun King attached to the idea of the unifying Cat.

The Egyptian Cat, Bastet was an ancient Solar and War Goddess in Egypt. The Cat Goddess was worshipped at least since the second dynasty. She was one of the favorite goddesses of the highly educated astronomically minded Amun-Ra priesthood. Her name means 'the female Devourer'! She started out being the protector of Lower Egypt (Magnetic North in our theory). Later, she achieved the titles *Lady of Flame* and Eye of Rá. Her revealing title of the Lady of Flame is very important in our theory, since we think that right here on the Ecliptic is where the largest and last of four fiery solar eruptions happened in the past causing the Magnetic Axis shift to begin.

This is how the book A Treasury of Classical Mythology by A.R. Hope Moncrieff writes about the tragedies of Phaeton, the Greek Son of the Sun.

While the swift-fingered Hours fitted on their clanking bits, and harnessed them to the chariot-pole, fond Phoebus anointed the youth with a sacred balm that would enable him better to bear the heat of his glowing course. Meanwhile shining Apollo plied Phaeton with warnings, to which his impatient son hardly gave ear.

"Keep heedfully the straight path marked by fearsome signs of beasts. Beware in going by the horns of the Bull and the mouth of the roaring Lion, and the far-stretched claws of the Scorpion or the Crab. ... Sink not too far down, lest the earth catch fire.

–A.R. Hope Moncrieff, A Treasury of Classical Mythology,

Thus, besides the ancient Egyptians, we noticed that the classical Greeks 2,500 years ago already thought the same Egyptian wisdom about the Bull and the Crab connected with the Fire from the Sun, which we see thousands of years earlier from the Egyptians.

Then as we read on further, we learn that in that solar disaster of the scorching Sun, even the powerful Ocean covering 70 percent of the surface of the Earth had to offer against.

Now they soared up towards the sky, so that the clouds began to smoke, and the Moon looked out with dismay to see her brother's car so strangely guided. Then turning downwards, as if to cool themselves in the ocean, they passed close over a high mountain, that in a moment burst into flames. Thus, fearsome disaster fell upon the earth. The Sun, instead of holding his stately beneficent course across the sky, seemed to rush down in wrath like a meteor, blasting the fair face of nature and the works of man. The grass withered; the crops were scorched away; the woods went up in fire and smoke; then beneath them the bare earth cracked and crumbled, and the blackened rocks burst asunder under the heat. The rivers dried up or fled back to their hidden fountains; the lakes began to boil; the very sea sank in its bed, and the fishes lay gasping on the shore, unless they could gain the depths whence Poseidon thrice raised his head and thrice plunged back into his shrinking waves, unable to bear the deadly glow.
 –A.R. Hope Moncrieff, A Treasury of Classical Mythology,

Due to the incredible force of that solar assault, the Keystone began to roll and created havoc on Earth. This well-hidden hot enigmatic holy Rock is worshipped all around the world amongst the Jews, Arabs, Christians and Pagans alike. That ancient Rock, that periodically determines the fate of our humanity, has secretly found its way into our varied faiths through the Rock that is housed in the octagonal shaped enclosure called The Dome of the Rock in Jerusalem. The beautiful Dome was completed by the Muslim Abd al-Malik around 692 AD. The Torah, the Old Testament, New Testament and the Koran contain enigmatic historical events tied to this superb Rock. It was where Moses preached and where Jesus was judged. The Muslims remember it as a holy site related to the Night Travel of Mohammed.

The nearby El-Aqsa Mosque and the Dome of the Rock served as the headquarters for the Knight Templars during the infamous time of the Crusades. The cube shaped Holy of Holies that was apparently part of the enigmatic Solomon's Temple became an important secret of the Masons, and likely hid the knowledge of cycles and timing with the

The World Tree   57

help of the Keystone rock. Naturally, more than useless cosmology, for most worshippers and pilgrims a mere physical touch of the Rock meant more than any secrets. Touching the Rock that secretly represented the Core of the globe became an act of paying homage to the gods for keeping us safe on this beautiful Earth.

This Rock was protected inside the Dome as the holiest rock on Earth to millions of believers of different Faiths of Grandfather Abraham. The famous Dome of the Rock is found in the holy city of Jerusalem, which is sacred to Jews, Muslims and Christians alike. One little known Jewish tradition claims that it is where the entire Creation began. Certainly, historically this is the rock where the faith of Abraham, father of Ishmael and Isaac was tested and it is where Abraham sacrificed a Ram in place of Isaac. This is also the site where the famous Temple of Solomon stood. Apparently, the temple was destroyed before 600 BC and King Herod later erected a Second Temple in its place. It is very likely that this 'Two Temple' symbolism was created by the Ancients to state that the First Temple of Mother Earth was destroyed in an axis shift and then after the stabilization of the foundation Rock, Second Temple was built in its place. The only significant remnant of that famed ancient Temple today is the westerly Wailing Wall, a special place of worship for the Jews. The trial and judgment of Jesus by Pilate was also connected to this Rock.

The ancient Jews who were stubborn and persistent in keeping their traditions and continued teaching the old religion, were readily persecuted by the Romans. To avoid the harassment, they found security in the caves of the mountains where they taught the ancient secrets. Their children set in front of the caves innocently, fooling the Roman soldiers when quietly playing with the pentagon shaped spinning tops. These four sided tops called Dreidel is a children's game of chance that is most commonly played around the time of Chanukkah.

Therefore, now we look at the Two Pillars cosmologically as a Dreidel like part of the Earth that is spinning in an ever increasing manner of a spiral path that eventually has to fall. Sensing that the Fate of Christ was somehow connected to this falling rock, we begin to peer at the cosmological clues where the Earth change scenario of the burning of the Earth began. What was Christ referring to in His saying about his second coming as a strong 'illumination'? Did He point toward a disaster approaching at the End Days involving the Sun eruption? We think He did, as in most Nativity paintings we detect a bright light from the dark midnight sky around the time of the birth of the Christ.

It is not a coincidence, that there are Two Donkey stars in the shaft of the Cancer star constellation. There is at least one Donkey present at Joseph and Mary's travel and the birth of Christ. This astronomical concept of the Cow and the Donkey is very well documented on famous paintings of various Nativity scenes by the painters of the Middle Ages.

Further clues, that point toward a disaster that happened at the nearby BEEHIVE Cluster is the followings. The Beehive Cluster M-44 in Latin is called Praesepe or 'Manger', from which apparently the Two Asses were eating hay out of. It is a linguistic enigma hidden from the public that it is the same 'manger' in the sky that was used to hold Baby Jesus. Now, it is not a coincidence that this celestial CRIB is spelled similar to CRAB in English. Thus, we can detect a design of the Cosmic Language in English. The words such as; Anger – M-anger – D-anger, likely referring to the Disaster of the End days to come with the beginning of the birth of the mythological Son of God.

The name of the astronomical Beehive **Praesepe** in Latin also bears both meanings for the Beehive and Manger. The similar sounding Latin root word **Praese** means Governor and **Praecessio** stands for the PRECESSION of the Equinoxes. Then from this same Latin root we find the title **Praetorium** that is translated in the Bible as the **Judgment Hall** where Jesus was condemned to be crucified.

"14 Now it was Caiaphas who advised the Jews that it was expedient that one man should die for the people. ...

... 28 Then they led Jesus from Caiaphas to the Praetorium, and it was early morning."

–Bible, John 18:14-28,

Since, the Hebrew and the Christian history lead us back from Rome to Egypt, we begin to search for the cosmic meanings and connotations to the BEES and BEEHIVES. Not only the Egyptian male fertility god MIN was known as "Master of the Wild Bees", but we find Artemis of

The World Tree 59

Figure 2-10: The Birth of the Descending Bee God of the Mayans. It represents the Cosmic Birth of the previous New Age. It is in the vicinity of the Cancer star constellation. Temple facing in Tulum, Yucatán, Mexico (Photo credit Wm. Gaspar)

Ephesus in present day Turkey, where the Mediterranean Goddess are plentifully ornamented with Bees carved on her legs on those ancient monuments. Also, according to Hindu mythology, when the Demon desired to establish his kingdom by driving out nature spirits - Devi, the Protector Goddess transformed herself into a Queen Bee and with the help of large swarms of bees covered the Earth with darkness. Similarly, the cosmic birth of the Mayan Diving or Descending God depicted on temple carvings in Tulum, Yucatan, Mexico is also shown in the Chilam Balam as a BEE God!

Thus, with the Bees and Beehives occupying main positions of transforming power, we are not surprised that the King of the Jews title can also be connected to the Beehive shape found inside the Star of King David, since Jesus apparently was derived from the Root of David. Even in the Old Testament of the Bible we find evidence that the Bees, the Fire and the LORD are connected in the same sentence.

> *12 They surrounded me like bees; they were quenched like a fire of thorns; for in the name of the LORD I will destroy them.*
> –Bible, Psalms 118:12

Another Beehive example from the Bible arrives in a hidden form in the Nativity paintings where the Donkey – that is one of the stars in the shaft of the Crab – is always present at the birth of Jesus. As we learned from Astronomy, the Beehive Cluster M-44 sits near the Ecliptic, the path of the Sun, just next to the Crab constellation. Therefore, if the Crab or the Bees are not mentioned then the Donkey will be assigned to the birth of the Son of God or the Son of the Sun. Out of the three options of Cancer / Crab, Bee or Beehive or the Donkey, definitely the four legged creature is the most hidden aspect of Astronomy.

Linguistically, the CANCER constellation is where the ancient bards marked the Earth changes that hit the globe ~12,000 years ago. Therefore, as a reminder for the bigger cosmic death, the CANCER word became the evil disease 'Cancer', such as lung cancer to remind us in our every day speech that Cancer was bad for us. The medical diagnostic word and the Mayan and other world mythology matched the 'old disaster' we were supposed to remember. Bad memories are

Figure 2-11: The Nativity painting by Edward B. Webster, 1956 Original is in the National Museum of American Art, Washington, D.C. One can detect on this modified drawing that the Light coming from the 5 pointed star in the sky and shines on Baby Jesus. The Light passes between TWO DONKEYS. Those two donkeys are found in the shaft of the CANCER star constellation and called Asellus Borealis and Asellus Australis. (Modified from E. B. Webster 'The Nativity' by Wm. Gaspar)

difficult to keep, thus the reminder had to be turned into a 'good news' as far as the myth makers were concerned.

This beam of light is the separating Torch between the East and the West powers, being essentially a half way marker between the two halves of the Full Precession. This cosmic divider is when everything happens in every mythology and creation legends. This beam of light is the fire that separates and unites. One is two and two is one. The Male joins the Female. The Yin and Yang of the positive and the negative attract and repel and will eventually unite here in a mind blowing cosmic union. This is cosmology above all the extreme galactic science of the spiral pair electro-magnetic forces interaction in the living waters of universe!

The history that we are learning – if it has not been clear by now – is no history at all. The history that was taught to us is nothing more than the mythological rendering of the Earth changes in various forms and fashions. Every godly hero and redeemer is assigned the same birthday, whether it was factual or fabricated. As it appears now from those old stories, the imams, rabbis, shamans, viziers and medicine men were mainly obsessed about how to maintain the old cosmic concepts that ripped the Earth apart. They worried about how to warn us about the arrival of the next huge Galactic Wave or the next Loud Voice of an Angry God. This myth making machinery that is in control of our creative imaginations and actual daily lives - is still in existence today. It is not as much the viziers or friars anymore who paint a blue sky with the angels of heaven, but it is the illuminated lords of the entertainment industry presenting the flying robots we have to battle against. Is there still a higher order of astronomy Priests who function above us, against us or for us? Who are they? Nobody knows!

Well, the Ancients were clear about their cosmology. The same cosmic Crab that we find with the classical Greeks is present in the creation tales of the Mayan Popol Vuh as the favorite food of Zipacna, the Crocodile / Dragon deity. The Crab is on the front cover of the bestselling book written by Maurice Cotterell titled *The Lost Tomb of Viracocha*, which deals with the Peruvians shamans. Does this world wide distribution of the mono-myth has its roots amongst the few survivals of the last Great Flood, those sages who might have been the remnants of the sinking continent of Atlantis?

Certainly, from the cosmology of the Egyptians we kept a

The World Tree 63

superstitious reverence for the Black Cat. We cannot allow a Black Cat to walk in front of us, because that means something bad will happen to us today! An encounter with the black cat is unlucky! Is it as unlucky as its relative feline, the saber-tooth tiger and the hairy mammoths that gone extinct at the time of the start of the Holocene Epoch 12,000 years ago. We know the mammoths did not die of hunger, since some of them in Siberia were found dead with food in their mouth. It had to be the swiftness of the Galactic Super Wave that rendered them extinct.

According to the research done by Allan and Delair in their book titled *Cataclysm!*, the bones of these animals were found tangled with other unrelated species on one side of the mountains. The assumption is that they were likely crushed by a very powerful wave of tsunami. Not just any average tsunami, but according to Allan and Delair, this huge wave of towering water deposited about 800 feet of sand from the ocean on only one the side of the mountains. That had to be a huge tsunami wave and likely proceeded by a magnitude 10 earthquake or maybe larger. There had to be little or no warning to it. Our ice age studies show that this time was about 11,650–11,680 years ago. Are we marching toward a similar, cataclysmic age?

We notice the **BEEHIVE** is in the middle of the sign. This Egyptian / Magyar sign marks the unification of the Two Pillars from Noon to One O'clock and then the well-known Star of David with its Beehive does the same on the Ecliptic area from Three to Four O'clock on our Celestial Clock.

The ancient Magyar language is closely related to the ancient Egyptian language in certain important cosmological words. In the Magyar tongue EGY means number ONE, but it is also the root word for sexual and any other kind of unification, just as the EGY sign means Unification of the Two Kingdoms of ancient EGY-pt. The unification between the old Sun King and Mother Earth as a Harlot, caused over 98 % of the Earth animals and humans to perish, thus it was a worthy unification by all races of humanity to remember.

While the Romans conquered most of Europe and spread Christianity, there were Native American Indians who equally well preserved a cosmic knowledge in a much simpler and easy to understand natural setting. The sacred rituals of the Native Americans helped us immensely to understand the four directions, the distinctive

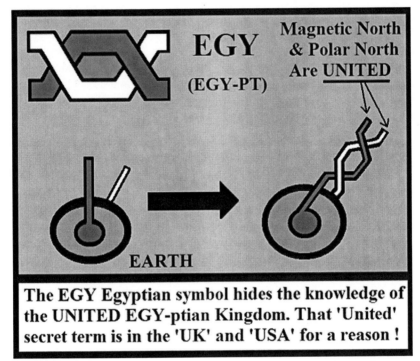

Figure 2-12: The EGY hieroglyphic sign in EGY-pt means the unifications of the TWO LANDS, which is nothing other than the secret cosmological meaning for the Unification of The Magnetic North Axis (Red) and the Polar North Axis (White) of the Earth's geo-dynamo. (Drawing by Wm. Gaspar)

colors and the forces of the Galaxy. Since we both live in neighboring state to Arizona, we also came in contact with the Navaho Nation whose legends we cherish.

For the Navaho (Dine'é) Nation, the star constellation Cygnus was immortalized as the TURKEY. The Wild Turkey who lived with the First people warned the First Woman and Man about the coming Flood. From Franc Johnson Newcomb's book, the Navaho Folk Tales - these creation stories are shown to be part of the same Mono-Myth that tells about the ancient Flood. The First World was called the Black World or the Place of Running Pitch. It might not be any different from the brimstone idea presented in the Bible.

The World Tree

This first world was a dark underground place where the only light came from burning pitch, and not many people lived there; mainly, it was the home of insects and crawling things. ...

"How can we leave this place?" inquired the ants. "Where will we go that will be any better? The rest of this world is burning and we cannot live there!"

Their Second World was better where they had a blue light coming through the sky, but that was the domain of the bird people, mainly. As expected, at the sight of the few people and the insects, a white crane arrived from the east and ordered them back.

Locust replied, "We cannot go back, for our land is on fire and will soon be gone."

Then the insects got to the Third World, or Yellow World, that was brighter than the previous one. It was the home of all kinds of animals and even humans. The life was good and abundant, so the animals and the humans over-populated the Third World. They soon all agreed that a council needed to be called to decide upon new laws to establish to divert disaster.

It is interesting and unusual to see that even the nature based Native American Indians would have the concept of overpopulation in their ancient legend. Was it at a time when there were many people living in Atlantis before the Flood?

Swift messengers were sent to ask all of the people of this Third World to attend a meeting where everyone would have a voice about how the new government should be formed. Firefly was sent to the north, Locust to the east, Honeybee to the south, and Dragonfly to the west. As these four flew in straight lines; they would never be confused about the right direction.

Before we go any further, we would like to mention the Honeybee and the South direction. The 'Beehive' is tied to the south direction, which is the direction of the axis shift in our theory.

In their creation legend the ancient Navaho council could not agree who to name world leader, thus the wise Owl suggested sending the four candidates to the edges of the world and whoever brought back something useful would be the leader. So they sent the Bluebird, Hummingbird, Wolf, and the Lion to their mission. Then they arrived to the pleasant Fourth World, where they experienced abundance. The animals and the people flourished and became more numerous. This is how the story continues:

The Flood

During the time the First People had been living in the pleasant Fourth World, they never thought that some day they would be obliged to leave, and with so little warning that most of their possessions would have to be abandoned. They little dreamed that they ever would be driven from their homes by the angry waters of a great flood.

During all these years of prosperity, Turkey had lived with the family of First Man and First Woman as a pampered pet. Sometimes he would take long Walks with First Woman's small son .One day, when Turkey and this small boy were walking along the edge of the mesa they heard a very queer noise coming from the north which sounded like "Sh-h-h-hiss, Sh-h-hhiss." It seemed like the angry voice of Grey Goose. It was the foam on top of waves of deep black water that rolled high above the land. As the first wave broke with an angry hiss, another just behind it marched forward to take its place.

In the above Navaho creation legend we notice the 'hissing' sound and the foam on top of the rushing water that eventually covered the whole land. It can certainly be a description of a tsunami that came quick and without warning.

In the Navaho creation legend they emphasize that there were 32 Navaho clans and **32 Council members.** That reminds me of **Jesus who lived to be 32-and-a-half years old.** Now, even more amazing is the fact that out of the 36 years that the Winter Solstice Sun travels through the Galactic Alignment in these last years between 1981 and 2016, the 32nd year is 2012!

One of the best cosmological ideas of the Mayans is their use of the spiral symbols. Also, we would like to demonstrate that the birth of the Sacred Child was associated with the Serpent at the dividing point between the Two Spirals. This was known to the Mayans and they clearly depicted it on the faces of their temples before the advent of Christianity.

The Hunab Ku Maya symbol is of major significance to our understanding of how the Forces of the Galactic Center work. First of

The World Tree 67

all, there are Two Spirals in a pair, just as the ones seen in the Galactic Center. Later, we will notice that the Spiral Pair reaches a ZERO POINT in the Center of the Pair, and that Time is what the Mayan Elders call Year Zero. If this model turns out to be correct, then we can expect to detect this same pattern in the movement of the axis of the Earth, and the Precession of the Equinoxes that we assume is recorded in the Ice Age cycles.

The second obvious piece of information from the Hunab Ku sign is that there are five parts to each Spiral of both, thus making it a Perfect Ten together in the Pair just as the Ten Fingers on both hands. Nature creates in five parts from one Spiral, and ten from a Spiral Pair.

The third part of the information we gain from the Hunab Ku sign

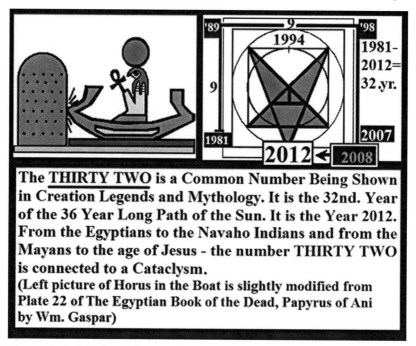

The **THIRTY TWO** is a Common Number Being Shown in **Creation Legends** and Mythology. It is the 32nd. Year of the 36 Year Long Path of the Sun. It is the Year 2012. From the Egyptians to the Navaho Indians and from the Mayans to the age of Jesus - the number THIRTY TWO is connected to a Cataclysm.
(Left picture of Horus in the Boat is slightly modified from Plate 22 of The Egyptian Book of the Dead, Papyrus of Ani by Wm. Gaspar)

Figure 2-13: The South Point of the Pentagram Star points to 2012. Also, the Hawk-headed Horus sits in the Solar Bark with the Sun on his head. His solar bark hits a Tombstone that has 32 stars. Does this point to a demise that begins in 2012-2013? Graph by Wm. Gaspar. Modified from The Egyptian Book of the Dead, Chronicle Books, Plate 22)

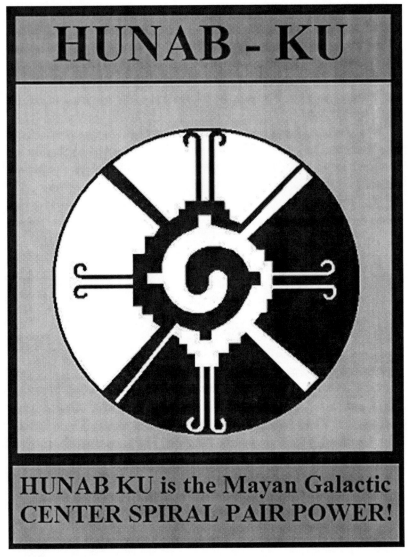

Figure 2-14: The Mayan symbol, HUNAB KU, is representative of the Galactic Center Spiral Pair Force. It is known as the Galactic Butterfly or the God of Measurement, Movement, and, for the ancient Mayas who apparently aligned themselves with the Galactic Heart's Super Massive Black Hole, The One God: that is, the Force.

The World Tree 69

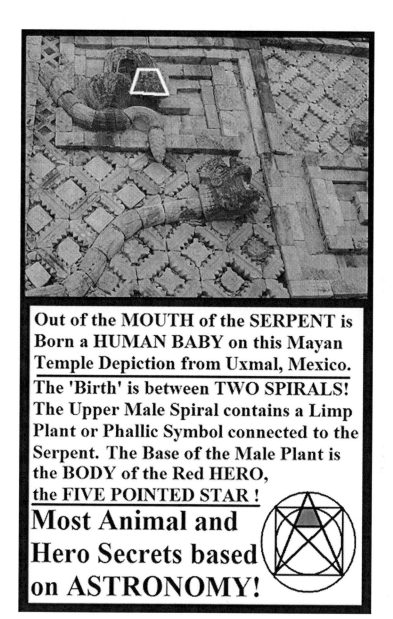

Figure 2-15: The Birth of the Sacred Baby out of the Mouth of the Serpent at the meeting point of the Two Spirals of our Galaxy. Temple Façade at Uxmal, Yucatán, Mexico (Photo credit Wm. Gaspar)

is that there are 36 separate outcroppings around the symbols with cosmic significance. The 36 years is half degree of Precession and the number of years the Solsticial or the Equinoctal Sun we count travels during the Galactic Alignment.

The Spiral electro-magnetic force that is the Creative Force in our Milky Way Galaxy breaks into five parts, according to spiral mathematics. That knowledge seemed to be in the possession of the Mayan astronomy kings, such as the Xiu family, who established Uxmal, in the Yucatan, in the year 751. The Xiu family still has living members of their clan flourishing in the Yucatan and near Uxmal. Don Gaspar Antonio Xiu is our friend and scholarly colleague living in the Yucatan Peninsula. He is the last living oldest adult descendant of the ancient Maya astronomy priests kings, whose family built the Uxmal Complex over 1,250 years ago, with clear and advanced knowledge of cosmology.

CHAPTER 3

From the Iron Rod of the Rock the Hawk Flies South

He destroys the blameless and the wicked . . . They pass like swift ships, like an eagle swooping on its prey.
—Bible, Job, 9

We ended our previous chapter with the birth of the Sacred Baby out of the mouth of the Serpent that emerges near the meeting point of the Two Spirals of the Galactic Bulge. This sacred and cosmologically very clever Mayan depiction was placed on the façade of the Temple in Uxmal, Mexico by the Mayan Astronomy Priests who founded the city around 751 AD. The mysterious Xiu Family of Mayan kings who erected the pyramids and those temple buildings, were naturally not Christians at that time, thus their Cosmological teaching is specific about a birth of a Sacred Baby tied to the Resonating Serpent of the Galactic Center Alignment. There is no intended and controversial ethnic history in this Mayan depiction, no Virgin Mother, no King Father, we only encounter a pure unadulterated symbolic language of Cosmology where the grandparents of the Baby out of the mouth of the birthing Serpent are the Two Spirals in the west. Cosmology just does not get any easier than that. On the Mexican currency and national flag we still find this same Hawk or Eagle who SWOOPS down to pick this Serpent out of a River – as the sacred representation of the axis shift.

We felt that the only meaningful translation we can provide to demonstrate our point is the most sacred alignment in the Milky Way Galaxy originating in the westerly Heart of the Goose and then further progressing toward the Swooping Hawk of Astronomy. Thus, in our first figure we align Mother Earth and the Sun as they would be seen from the Galactic Heart of the Goose in 2012 - 2013.

We would like to introduce the mythological parts of the Geo-

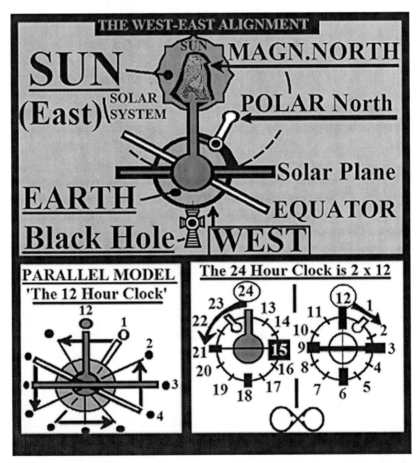

Figure 3-1: The East West Alignment of the Sun, Earth and the Black Hole on the Bottom contains the Serpent, Harp, Hawk, Keystone Rock of the Earth as equally important representatives of the Magnetic North. (Drawing by Wm. Gaspar)

The World Tree

dynamo that we presented in our Cosmic Model. The first and the most important animal is the Hawk in the position of the Magnetic North. The star Vega ('Valley' in Latin) is the official position for the Magnetic North, and it is known by its older name as the Swooping Hawk - the main star of the Harp / Lyre star constellation. This famous Hawk supposed to 'swoop' to the Throne of Isis, which is the star constellation Cassiopeia. It is identified by the Polar North, because in the middle of the month of the Virgin, in September the Throne of Cassiopeia is next to the Polar North. Thus, when the young Prince Hawk, the Egyptian Horus will swoop to the Throne of Isis, He will become the Son of the Sun, SA-RA.

This West to East alignment demonstrates why the Magnetic North is represented by so many different concepts, such as the Rock, that is the Core of the Earth, the Iron Rod, the Hawk and the Harp. Further on - because the Sun eruption happens during this alignment – also the Sun and the Serpent that represents the Sun eruption are tied to the Hawk, thus Horus is commonly depicted by a Hawk with a Sun on its head surrounded by a Serpent ready to strike. Therefore, we begin to conceptualize that on the stages of international mythologies, the Hindu Snake charmer, or even Šíva, the Hindu Sun god whose third eye erupts from the center of his forehead, is nothing different than the Egyptian Horus or the Mayan Baby out of the Serpents mouth or even our heroes from more recent European tales.

In the picture shown on the next page, the Blue Heron/Phoenix /Stork is shown so that we know that it is about a new birth. The Stork brings the baby. The Old Sun King died; welcome the new Hawk Prince!

This is how the Egyptian, Mayan, Hebrew, Roman and other astronomy priests of the Order of the Sun God RÁ / RÉ decided to portray the sudden Magnetic North Axis shift to the Polar North area: with the sudden swooping down of a hawk on a prey. By "sudden," we mean a few weeks to a few months for the first larger movement in one direction. Certainly we are not expecting a comfortable, slow, gradual shift over years or decades! One would not get a Swooping Hawk to represent a slow axis shift. The whole process will take decades before the axis stabilizes, but the first shift of the axis will happen fast and unexpected. This is how the famous E.A. Wallis Budge, one of the foremost Egyptologist wrote about the Egyptian

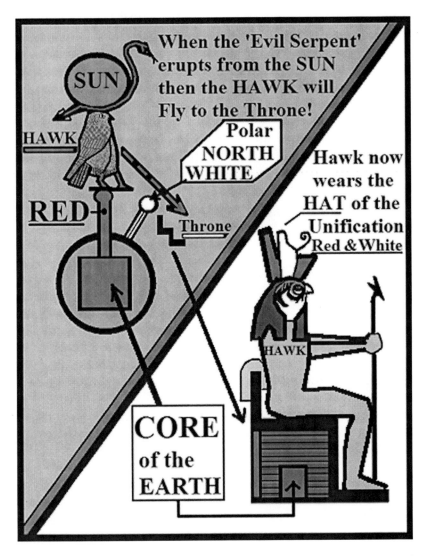

Figure 3-2: The Horus Hawk is commonly seen wearing the Sun on its Head and around the Sun is a Serpent coiled up ready to strike as the flying feathered serpent of a Sun eruption would. It is also seen as the Prince sitting and when the Red Brick shaped Core is shifted to under the Throne (Picture by Wm. Gaspar)

The World Tree

Figure 3-3: The Two Hawks on either side of the dead Old Sun King represent the magnetic North and the Polar North. The headdresses of the hawks tell the exact Cosmology. The throne on the head of the Hawk means that the Magnetic North shifted in the position of the Polar North (Vignettes from the Book of the Dead, Chapter 17- Drawing by William Gaspar.)

Astronomy Priests of RÁ.

> IX. *Now, the kings of Egypt were always chosen either out of the soldiery or priesthood, the former order being honoured and respected for its valour, and the latter for its wisdom. If the choice fell upon a soldier, he was immediately initiated into the order of priests, and by them instructed in their abstruse and hidden philosophy, a philosophy for the most part involved in fable and allegory, and exhibiting only dark hints and obscure resemblances of the truth.*
> –E.A. Wallis Budge, *Legends of the Egyptian Gods*

Figure 3-4: The storks and herons stand on ONE LEG in the water that is related to the birth of the Sun Prince and the following watery situation. (Paintings by the US / Chilean artist, Gaston Kessra, a student of mysteries)

Thus, every animal and hero represents a well planned cosmic secret propagated by the mysterious Astronomy Priests of Rá a Sacred Order that is still hidden and alive today amongst us. According to their ancient cosmic teachings there was a very exact meaning to the hawk, ram, goat, crab, serpent and even the stork. The stork standing in water on one leg represented a very specific concept during the earth change years.

So, why do we see a blue heron or a stork in these pictures? I am sure everybody knows that the Stork brings the Baby, even the sacred baby of the Old Sun King. What is so characteristic about the Storks to be in charge of delivering babies, especially the Son of the Sun? Well, where do storks nest? In some parts of the countryside in Europe, we can still see many storks nesting on the CHIMNEYS of old adobe houses! So mythologically speaking - why pick a bird that nests on the chimney? Not to mention that any bird that nests on a chimney would

The World Tree

block Santa's way into the house to bring presents. We think that the stork, with his habit of nesting on the smoky chimney, represents the burning that happened after the birth of the Son of the Sun, which is a secret reference to the huge solar eruption.

The biggest Sun spot eruptions of the past happened first at the Goat star Capella. Then the fourth and most powerful eruption happened when the Sun was at the Beehive cluster next to the Cancer star constellation. After the Cancer, the next star constellation on the ecliptic of the Sun is

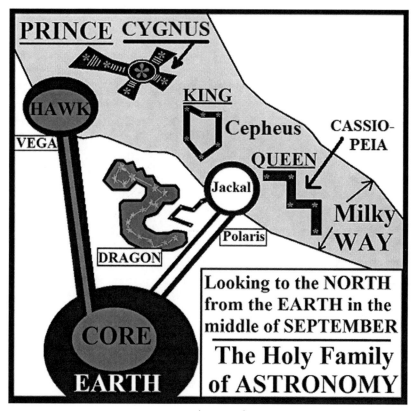

Figure 3-5: The Swooping Hawk / Vega / Magnetic North Flies to the Throne of Cassiopeia. This is the basis for the innumerable Paintings and Statues displaying Horus, The Hawk-headed Prince sitting down on the Throne of Cassiopeia, who is the Egyptian Queen Isis. (Slightly modified from Chet Raymo, 365 Starry Nights Drawing by Wm. Gaspar)

Leo, the LION. So, the Priests of Rá decided to put the armless corpse of the Old Sun King— the Pharaoh—on a bench that looks like the legs of a lion. Certainly, the lion has become an impressive cosmological symbol in most part of the world. The lion is the end stage of an axis shift that begins with the concept of the Hawk flying south.

Egyptian mythology commonly displays these Three together as the holy family. The Egyptians made it very clear that they talk about these three. They even placed the Sun over the Goose's back, which is the hieroglyphic sign, called SÁ-RÁ / Sháh-RÁ (SaRah), the Son of the Sun! Thus, to say that the Sun had an eruption and born a Male Child, the Egyptian Astronomy Priests used the SÁ-RÁ hieroglyphic sign, but what is interesting about this that they employed the West Power (Sá)

Figure 3-6: The Hawk in Egypt wears the Throne on His Head and the Hat of the United Kingdom also portrays the same meaning of the cosmology of the Magnetic North shifted to the Polar North (Drawing by Wm. Gaspar.)

The World Tree

Figure 3-7: The Holy Family of Cosmology. The Prince, the Goose 'SA' is shown with a Sun 'RA' on his back. The Father has the DEL-Ta shape and the Ankh represents the Queen Cassiopeia / Throne. (Drawing by Wm. Gaspar)

and the East Power (Rá) to emphasize the Alignment involved in that.

So, our question is this: - 'Was Dan Brown's SARAH, a daughter born to Jesus, the real secret, or what we are presenting tells a more reliable astronomical, mythological and scientific tale? We will let the readers decide.

The DELTA male force and the ANKH female force both had to be the domain of the Globe undergoing earth changes. We started sensing that it is where we are going to find the key to Cosmic Sex. Now, the RÉ / RÁ priests of astronomy decided to place the Core of the Earth under the Throne, so one would know that the Core of the Earth has shifted to the Polar North. **THE RED BRICK UNDER THE THRONE IS one of the most obvious SECRETS OF THE AXIS SHIFT!** The CORE of the Earth was depicted as a RED BRICK, a CUBE – just as what we see with the KA'ABA (Cube) stone

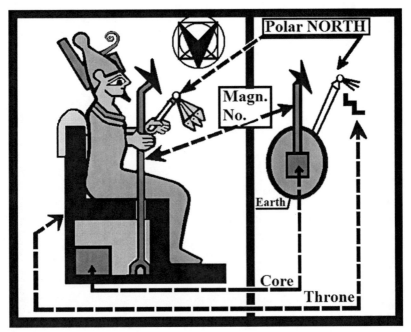

Figure 3-8: The Earth's Core, (the kiln-fired Red Brick) is shifted the Throne of Isis: that is Cassiopeia. The Iron Rod is called 'VAS' or 'Was' in Egyptian (Drawing by Wm. Gaspar)

The World Tree

of Mecca. The Red Brick is the Cube and it is also the Keystone body of Hercules. When one examines the title of Hercules as the 'kneeling giant', some of the religious practices of Catholics and the Muslims might be explained. We think that most religions teach and believe in an Almighty God. The differences of various religions seem to be partly in their special understandings of the unexplainable and partly in their physical practices and rituals involved during praying. This latter activity is what differentiates us from each other, while the Cosmology unites us all under one flag.

The pharaoh is holding the Vas Rod or Scepter, the sign of rulership in the position of the Magnetic North where it was before the Axis Shift, but here the Core has already shifted. Interestingly, the 'Vas' word in Magyar means iron, just as the iron core of the Earth. A lot of the Magyar and the Latin words are very similar or even identical with the Koptik Egyptian words and that helped us solve the riddle! Words or expressions such as Vassal (bondman / subject / ruled) are probably derived from 'VAS' as the concept of the Iron Rod of the Magnetic North attacking and ruling the Polar North.

The name of Nordic Valkyries ('VA') may also hide the knowledge of this 'VA' related secret. They were evil beings first, some sort of angels of death, who guided the heroes to the great hall of Valhalla (another 'VA'). On the battle field they were Odin's virgins who flew in the form of the Swans above the heroes. Andrew Collins book titled The Cygnus Mystery has a collection of examples about the role of the Swan and Goose in ancient ritual and shamanic depictions that lead us back to the important Cosmic role of the Swan / Goose in the endless tales around the World. With the Eagle or Hawk sitting on her right knee before all creation began. Does she represent the Galactic Hearts CYGNUS / Cross – she has the Dark Rift for the Cosmic Womb in the endless cosmic ocean.

Thus, we find the Swan / Goose / Northern Cross now as the Galactic Creator Mother in the neighborhood of the Dark Rift with the Eagle or the Swooping Hawk by her knees. Thus, we are interested if the Finn – Ugor people understood the Galactic Heart Force from the west. We sense, that besides the Egyptian, Mayan, Arab, Judeo-Christian and Magyar Cosmology we will find clear cosmic patterns in the Creation Legends of the Finnish, Vikings, Germanic, Keltic, Aboriginal, Native

Figure 3-9: The Daughter of Nature, Luonnotar is the Creator Goddess of the pre-Christian Finnish People. The Ancient Finn-Ugor legends are important secrets. Here Luonnotar represents the westerly Super Sun of the Galactic Bulge, where the Dark Rift is positioned between her Two Legs on the Astronomy chart. (Slightly modified from Arthur Cotterell's book the Norse Mythology by Wm. Gaspar)

The World Tree 83

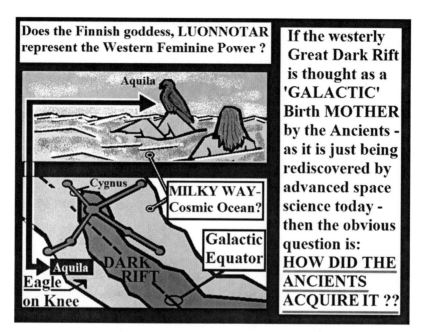

Figure 3-10: In the Galactic Center of the Milky Way Galaxy we find the Creator Mother – who we know in Astronomy as the Swan / Goose / Northern Cross. Lying in the west and her between her Long Legs - that point toward the east, toward the Sun King – is the Dark Rift, mythologically known to us as the edge of the birthing place of stars in the Super Massive Black Hole. (Drawing by Wm. Gaspar)

American and African Tribes.

Therefore, the Swooping Eagle or Hawk as we observed so far is not only the intellectual domain of the Egyptians, Persian, Chinese, Native Americans, but we begin to start seeing them also with the pre-Christian nomadic White People. Naturally, for most of us the information is provided in the Torah or the Old Testament of the Bible. In the book of Job, one can find a quote relating to the Swooping Eagle flying south.

"*9 He made the Bear, Orion, and the Pleiades, and the chambers of the south ...*

25 "Now my days are swifter than a runner; they flee away, they see no good.
26 They pass by like swift ships, like an eagle swooping on its prey."
–Bible, book of Job 9:9-26,

A little further on in the enigmatic Book of Job we find further references to the Chamber of the South.

"9 From the chamber of the south comes the worldwind, and cold from the scattering winds of the north."
"17 Why are your garments hot, when He quiets the earth by the south wind?"
–Bible, book of Job 37:9-17,

The Swooping Hawk flies south. Another referral to the axis shift toward the south is the mentioning of the South Chamber. The Chamber of the South is represented by the astronomical symbols shown by the Mithraic Mysteries Bull Killing scene that is secretly still kept alive in the Vatican. Thus, there are clear astronomical clues even in the New Testament that reach back to the teachings of the pre-Christian Mithraism. An excellent and astronomically revealing book on the subject is 'The Origin of the Mithraic Mysteries' by the excellent Professor David Ulansey who is one of the best authorities on the subject. The Chamber of the South from the Old Testament – as we will clearly demonstrate later - is the same cosmological concept as the Hawk Flying South or the Whale of Jonah closing its gaping mouth.

As the Hawk swoops down on its prey, that is how the Magnetic North Axis of the Earth swooped down on the Polar North in the past, when our globe's geo-dynamo went through its regularly scheduled maintenance of recharging the failing magnetic field. We think that one of the branches of our World Tree — in this case, the Magnetic North Tree is connected with the Tree of Knowledge, and then the Polar North Tree has to be connected with the Tree of Life, mythologically speaking.

This ancient tragedy that was followed by a huge worldwide Flood is what the astronomy priests of the Egyptian and Mayans recorded in mythology and related religious writings. It happened during the time when the Sun in his ecliptic was crossing and aligning with the cave-shaped **Galactic Dark Rift** that the Mayans called **XIBALBA BE and is thought to be the Cosmic Womb.** This path of the Sun lasts about 36 years, and we assume that it started around 1981 and will end around the year 2016.

Naturally, each culture, society and civilization favored a different calendar or a different musical instrument to tell that specific time when from the Resonation from the Powerful Word of God or by a sacred Arrow from the gods began the celestial music. The Word and the Music definitely became all of a sudden louder and disrupted the harmony of peace and the world descended into chaos. The flute, the drum or any stringed instrument would refer to the music that happens to the Magnetic North when God's resonation, huge lightening and fire will arrive to destroy nations.

13 For I have bent Judah, My bow, fitted the bow with Ephraim, and raised up your sons, O Zion, against your sons, O Greece, and made you like the sword of a mighty man.

14 Then the LORD ('ADON' in Hebrew-auth.) will be seen over them, and His arrow will go forth like lightning. The Lord God will blow the trumpet, and go with whirlwinds from the south.

–Bible, Zechariah 9:13-14,

Thus, the Greek Harp, the Biblical Trumpet, the American Indian Kokopelli's Flute, or an African Shaman's Drum is always there to remind us that the pleasant sounds of the Harmonic Resonation from the West and the East can all of a sudden change to a loud shriek on the command of our Creator. When the music changes, or at the time when the peaceful quietness of the early morning army encampment at once interrupted be the bone chilling squeaking sound of the loud trumpet – God or Nature is announcing that the Global War has begun!

Examining the more classical instruments such as the Piano or the Harp, we find that the Harp / Lyre (LYRA) star constellation is that the only musical instrument amongst the 88 astronomical star constellations. The number 8 symbolizes eternity, and the number 88 may refer to both halves of the Precession of the Equinoxes repeating in cycles over and over again. The number 88 shows up on the **Keypad of modern PIANOS**. There are 88 keys on the keypad of the modern piano, just as 88 star constellations in the sky. The other obviously astronomical and cosmically harmonic numbers also show up in the mathematics of rhythms and music. Some of these numbers are the 5, 7, 8, 12, 36, and 52.

Of the **88 piano keys,** we recognize **36 black ones** that correlate to the 36 years that is about the size of the Sun in the sky, and also the 36 years that equal One Half degree of Precession.

Furthermore, out of the 88 total keys on the piano we subtract the 36 black ones: and that leaves us with **52 white keys.** The **52 white keys of the piano, then, agree with the 52 years of the conjunction of the Sun and the Moon**, the Mayan Year of the Harmonic Convergence.

Then there are 7 octaves and 12 parts in each octave. All of these numbers strangely remind us of the natural calendar rhythms. Well, we felt that a musical departure into cosmology will be refreshing.

Returning to the Hawk of the Core of the Earth, we wonder why the Greek Hero Hera-kles, the Roman Hercules, the English HERO, Magyar (h)ER, Egyptian HERU, and the Latinized HORUS name are denoting the POWER of the Swooping Hawk or the Swooping Eagle or more specifically the Power that turns the Rock, the Keystone of the Earth. These heroes point to the Magnetic North as the Heroic part of the Earth's Geo-dynamo.

Can we tie the Galactic Womb from the west and the star Vega, the Swooping Hawk to the north of Mother Earth, to the famous Hercules together astronomically? Thus, mythologically then – can we also connect the Philosopher's STONE or the famous KEYSTONE of the Bible to HERCULES? The answer is YES! The ancients did not want to leave any stone unturned or any doubts in our minds about their message of astronomy, thus they tied them all together! They wanted the seeker to feel that the secret key that they discovered will open all the doors of the Chamber of Cosmological Mysteries

We see that the constellation Hercules, according to Chet Raymo, is just right below Vega, the Swooping Hawk that identifies the Core of the Earth. We can see that Hercules became part of the mythology related

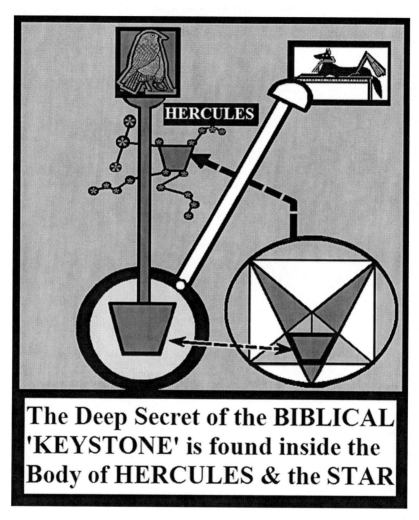

Figure 3-11: The BODY of HERCULES is called KEYSTONE in astronomy and one can look in the Bible to find the significance of the magical Keystone. The deltoid shape of the body of Hercules itself contains major enigmas. Thus, if a Female is the 'VA', hidden in the upside down 5 pointed Star, - then we expect the Male to be the upright Red 5 pointed star, the 'Red Hero'. (Drawing by Wm. Gaspar)

Figure 3-12: The Serpent has eighteen waves under the Solar Bark of the Sun that counts to be 36 when we look both ways. Also, pointing out the phallic symbol as the 'Temple' of the Face of the Egyptian God leaves us with an obvious Eruption shown next to the Sun. (Modified from the Tomb of Ramsses IX by Wm. Gaspar)

to the Core of the Earth, and thus, the Magnetic Axis Shift Secret of the Geo-dynamo. Linguistically, we begin to see an obvious and remarkable relation between HERU (the Egyptian Hawk 'Horus') and the HERO of the Greco-Roman mythologies. Thus, do all human Heroes of the creation legends originated from this most ancient Egyptian legend? We even start peering at the claimed-by-all capital of the Judeo-Christian / Arab city of JERU-Salem – in which the sacred Rock is kept in a Dome

The World Tree

— we notice that if JERU was written in the Spanish language then it would be pronounced as the Egyptian HERU and Salem means 'Peace' in Arabic! Thus, this holy city, which has changed names and ruling lords from the time of Abraham is was fought over by every organized religion to own the Rock of God. The astronomical significance always

Figure 3-13: The Hawk on the Throne with the Sun and the Serpent. The brick under the Throne is the Core of the Earth. In the upper left-hand corner one can notice the ALL-SEEING SACRED EYE, the symbol for the Galactic Black Hole (as a black pupil). (Solar Barque in the Underworld, Papyrus of Herytwebkhet Drawing by William Gaspar.)

was known to the Priesthood and was available for the few initiated, then the ugly place of bloody war and ethnic hatred remained the lot for the rest?

The number 36 is the number of years the solstice Sun travels across the Galactic Bulge every about 6,000 years. The Egyptian pharaonic name, **RAMSES (RA-MOSE-S) means, 'the SUN BORE HIM'!** (This refers to the birth of the Son of the Sun.)

Linguistically, RA would be the easterly 'Sun' and then MOS would be the 'birth' of the Son of the Sun traveling to Mother Earth unifying the Two Pillars. The Egyptian letter 'S' at the end of the name written as a Candy Cane (='S') the Galactic Plane force that turns the Earth from the west to the east. This Egyptian Hieroglyphic Alphabet is a little more complicated to most readers, but we mention it, because it is still a very active foundation to the international Sacred Cosmic Language.

The Dog Star, SIRIUS (means = scorching, in ancient Greek), has an about 1,460-year cycle, and 4 dogs (4 x 1,460 = 5840) approximate the quarter cycles of the Precession of the Equinoxes found in the Ice Age cycles. Thus, 8 dogs would be (8 x 1,460 = 11,680), 11,680 years, almost exactly the half cycle of Precession. This is the time frame shown in the Vostok Ice Age cores when a major change happened repeatedly. The Egyptian art, when read properly, can yield important cosmological knowledge.

In our theory, this is the mythological unification of the Two Kingdoms of ancient Egypt. Now, during this unification process, the King called CEPHEUS, which means HEAD in Latin, not only says that he is the Head of the Kingdom, but also to refers to the 'Beheading' process of the Old King (Sun /Magnetic North axis). The Old Sun King must go through this beheading to achieve the intended result of allowing a new Sun Prince, Horus, to be born. With the rebirth of the new sun, the recharging of the Earth's magnetic field is achieved and the protective forces can be reenergized. This travel and return of the Magnetic North axis is also known as "traveling to foreign land," or the captivity of the Jews in Egypt, a difficult shift from history to cosmology.

The above Mayan picture demonstrates that, from the Arab astronomers to the Mayan priests, the Message is Uniform and it is Astronomical! With a uniform astronomical message, one can now precisely assemble the Cosmological tale of the Universal Mono-Myth of periodic cataclysms!

The World Tree

Figure 3-14: The Mayan beheading, shown at the ball court of Chichen Itza. Notice the similar form of the Kneeling Man (Hercules), who is beheaded in Figure 3-11. (Photo credit by Eva Gaspar)}

CHAPTER 4

Venus, Darkness and Beheading

> *Come, I will show you the judgment of the great harlot who sits on many waters, With whom the kings of the earth committed fornication, and the inhabitants of the earth were made drunk with the wine of her fornication.*
> —Bible, Revelation 17:1, 2

The concept of 'beheading' is not only a Mayan concept, but it is encountered in every corner of the Earth. Its mythological examples are numerous and the cosmological meaning represents the Axis shift of Earth from the Magnetic North to the Polar North.

The ancient astronomy priests were very concerned about the Beheading of the Earth and therefore with the natural rhythms of the PLANET VENUS that they used to count the long-term time keeping. Planet Venus, along with Mother Earth, is known in mythology as the goddess of beauty, and is an integral part of most creation stories and religious tales involving princesses, queens, the Harlot, and the Virgin. What appeared so unique and important about the Venus cycles for the astronomers? What were they counting so meticulously in the cycles of the planet Venus? Could the astronomers foretell the ending of an era just by knowing the cosmological characters of our sister planet? We believe so.

The ancients counted 96 pairs of Venus Transits in front of the Sun, at about 121.5 years apart. At the last pair, they would mark the

The World Tree 93

end of a Half Precession of the Equinoxes. When they arrived to the end of that Portal – which we assume begins with the year 2012 – then first with the beginning of the larger earth changes, the Harlot would entice the old Sun King to erupt prematurely then the birth of the Son of the Sun would began a process of an axis shift that would close the gaping legs of the Harlot Mother Earth and make her to be like a Virgin. Following that, the erupting underwater ocean would turn the waters of the oceans blood colored and the erupting land volcanoes of the Fifth Hour of the Book of the Dead would create the Lake of Fire and the subsequent darkening of the sky. From the 4th and the 5th hours of the extreme hot, next year we entered a year that was so cold that people barely could imagine. Then it got very dark from the metallic dust of the volcanism. The ancient classical Greeks maintained a story of this period as the flight of the metallic Stymphalian birds that darkened the entire sky. The Darkness lasted about Three and a Half years. How dark was it? At least as dark as what happened in the 6th Hour of Christ on the Cross.

33 Now when the sixth hour had come, there was darkness over the whole land until the ninth hour.
                                    –Bible, book of Mark 15:33,

Naturally, if we believe that there was a universal Mono-myth that has been derived from at least as far as the sinking of Atlantis, then we hope to find evidence of the Secret Formula of marking these teachings down as the Last 12 years of the Hero's Journey before Death and Flood arrive. Now, we can check back before the time of pre-Christian Egypt, Babylonia, and the classical Greco-Roman times to find similar timelines recorded. We investigate to see if there was darkness during the 6th to the 9th years of the Labors of Hercules. Certainly, we expect to find similar allegories in the stories of the last Twelve Hours of the Egyptian Book of the Dead and in the Last Twelve years of the Labors of Hercules. Furthermore, in the Twelve tablets of the Babylonian Gilgamesh and again in the Twelve Hours on the Cross in the Sacrifice of Christ – then we can begin to collect the astronomical sequences of events the ancient seers, prophets and sons of gods brought us to remember. Therefore, we examine the Sixth Labor of Hercules.

> *Hercules next journeyed off to Crete to accomplish his sixth task, the capture of a mad bull given by Neptune to* **Minos**, *king of the island. The god had sent the animal with directions that he should be offered up in a sacrifice; but Minos, charmed with his unusual size and beauty, resolved to keep him, and substituted a bull from his own herds for the religious ceremony. Angry at seeing his express command so wantonly disobeyed, Neptune maddened the bull, which rushed wildly all over the island, causing great damage.*
> –H.A. Guerber, The Myths of Greece and Rome

Although, we have not noticed any darkness in Guerber's rendition that anyway sounded like the Seventh Labor in other versions, we registered the great damage and a reference to the Ocean in the form of Neptune. One thing we did identify as part of the Mono-myth from ancient Egypt was the name of the king MIN-os! We have seen that name for the Egyptian God of Male Fertility and Rain whose name was MIN! Not only the Greek Minos was charmed by a beautiful BULL, but we remember that the sacred animal of the Egyptian Min was a White Bull. Any bovine being white and bigger than that can only be the White Buffalo Maiden of the Lakota Indians.

We returned to the Sixth Labors of Hercules to see what we could find in the way of darkness that will be resolving in the Ninth Labor.

> *The sixth labour was hunting out the Stymphalides, those same arrow-feathered birds of prey that troubled the voyage of the Argonauts. Lake Stymphalis in Arcadia was their breeding place, which Hercules found black with such a throng of the mischievous fowl that he knew not how to deal with them.*
> –A. R. Hope Moncrieff, A Treasury of Classical Mythology,

Then we find again a few words about the metal clawed Stymphalian birds and the stagnant waters of the Lake in Guerber's book.

> *The Stymphalian Birds*
> *Eurystheus, well pleased with the manner in which Hercules*

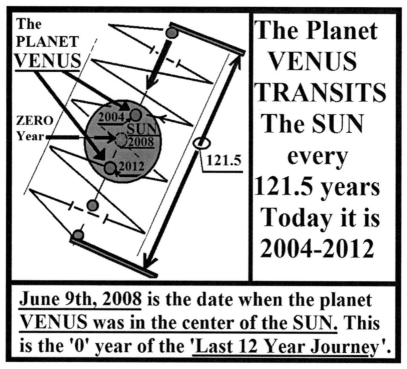

Figure 4-1: The Pattern of the planet Venus observed from Mother Earth is shown to yield a 121.5 year cycle. This was part of the secret counting measure of the ancient Astronomy Priests that remained well hidden from the public (Drawing by Wm. Gaspar)

had accomplished eight out of the twelve tasks, bade him no go forth and slay the dangerous, brazen-clawed birds which hovered over the stagnant waters of Lake Stymphalus.
　　　　　　　　–H.A. Guerber, *The Myths of Greece and Rome*,

　　Thus, the stinky stagnant waters from the Lake from before had resolved by the Ninth Labor. The Darkness resolved when the Mother Earth, now with legs tied together as a Virgin's, experienced another birth of the Son of the old Sun King. Now, who would want to find a woman greater than Mother Earth? Would any earthly flesh woman

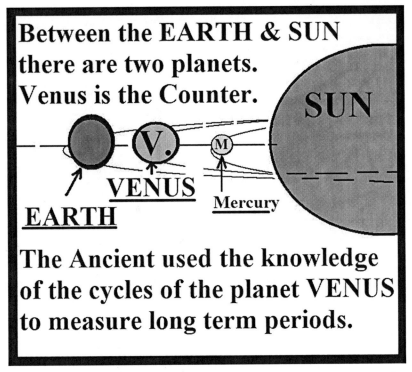

Figure 4-2: The Planet Venus is the second planet from our Sun and is the best counting measure for long-term cycles of thousands of years. (Drawing by Wm. Gaspar)

– even when sanctioned by a deity - could be a more important female then our own Mother Earth, who provides us with everything on the physical level that we need to live? We would have a hard time finding any greater Harlot or Virgin than our Mother Earth.

Every 121.5 years, Venus would pass in front of the Sun, as it is seen from the Earth. Mostly, it happens in pairs, eight years apart, but there are times with only single transit, if the first easterly meeting between Venus and the Sun is further down than the edge of the Sun. **This last 96th VENUS TRANSIT PAIR that we are experiencing began on June 8, 2004, which, in the Mayan calendar, was the day 6 WIND kin 162. The transit**

The World Tree

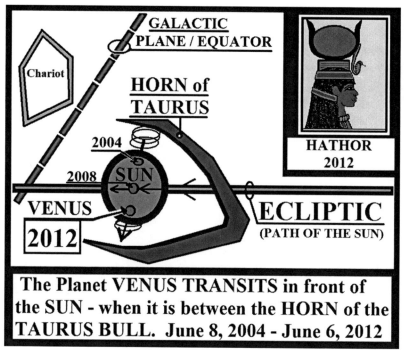

Figure 4-3: The VENUS TRANSITS of 2004 and 2012 in front of the SUN when the Sun is between the Horns of the Bull, Taurus (Drawing by Wm. Gaspar)

on June 6, 2012, will be the day 1 WIND kin 222. This is the Portal, the heavenly gate that will open up to us by the hinges of the year 2012 to see the rebirth of a New Age.

Will it be a dimensional change, physical destruction, enlightenment or all of the above? A very important date in the Center of all of the Venus Transit Pairs is halfway around the Sun, when the planet Venus is practically in dead center of the Sun in the West, looking at it from our Earth. Venus is then behind the Sun. This most important station and date in her 121.5 year journey about the Sun is the time when she is IN THE MIDDLE OF THE SUN, behind it! That is the year Zero for the 121.5 years journey, thus the remainder ~120 years is marked in religious writings as 'God gave men 120 years to live'.

One of the most famous and vocal proponent of the Sun induced

Earth changes is our good friend and colleague Mitch Battros, whose book titled Solar Rain is one of a kind theory being proven by the emerging solar science. His web site about the earth changes is an excellent source of daily news of earth change weather. He readily joined us on our 2006 Taos Earth Changes conference with John Major Jenkins and he also came to San Diego in May 2008 for the Between Two Worlds conference where Dr. Jaramillo, Dr. Gaspar, the descendants of Mayan kings Don Gaspar Antonio Xiu and Mitch Battros were the main speakers. We attempted to raise awareness to the approaching possible earth devastations awaiting our changing globe. He is relentlessly interviewing Mayan Elders alongside the European Space Agency, NASA and NOAA scientists to bring a scientific understanding to the complex issues of Sun directed earth changes and the parallel wisdom of ancient holy books.

Hathor is commonly shown with the Taurian Bull Horn and with the Sun between the Horns. This can represent 2004 or 2012, but today we believe that it will be marking the infamous year of 2012 and the beginnings of the more severe Earth changes.

Thus, the three important years connected to the Venus Transit Pairs are the years 2004, then 2008 and the well-known year of 2012. In classical Greek mythology, the Three Muses are shown around the Sacred Tree with a Serpent around it. All the way across the continent, in Egypt, it is slightly more difficult to differentiate between the Three Goddesses. The most well-known trio of females is Hathor, the sensual goddess of love, Isis, the perfect wife, and Mut, the elderly female, the widow. Naturally, this late in the game we certainly think that the unlucky female who is being torched as much as the widow of a Hindu Maharaja - would be the Old Widow, the Egyptian Mut who is usually shown dressed in dark apparel. Again, this revives the international Cult of the Black Virgin adored by many throughout the ages. Whether it is the Harlot, the Virgin or the poor Widow, the burning from the Sun is still happening to only old Mother Earth.

Just recently, Spaceweather.com, operated by NASA and NOAA, reported that **the SUN was growing "MICKEY MOUSE" EARS**. Naturally, it could only be detected by certain sensitive instruments, not with the naked eye. The interesting fact of the radiation making the ears on the round sun also made the sun appear more 'square':

The World Tree

SQUARING OF THE CIRCLE! The squaring of the Sun was happening in all four quadrants, and those who are interested in it can find it in the archives of the Spaceweather web pages. Shortly after the Mickey Mouse Ears, we started hearing so much more about the vivid blue NOCTILUCENT clouds that appeared electric in their shape as Blue Great Whales floating through the sky.

A few northern European observers reported that they began to see wrinkled noctilucent clouds in the sky that resembled the skin of the **Great Blue Whale.** Both of the above facts were a bit scary, because we started thinking back to the ancient mythologies when the Whale swallowed Jonah and also the stories about the Old King (Sun!), such

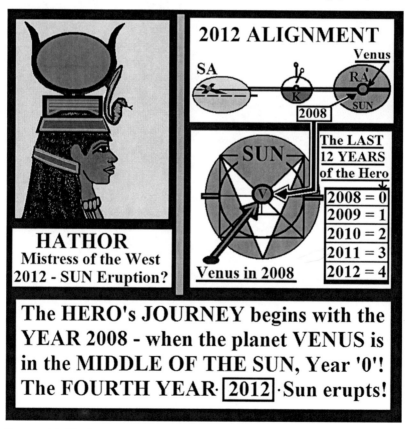

Figure 4-4: Hathor, Mistress of the West (Drawing by William Gaspar.

as Midas and others who grew **BIG EARS**. Certainly, the inference was usually in the form of a contestant getting blessed with Dumbo ears after losing a contest to Apollo, the Sun King. Are we loosing the contest to the old Sun King and began to see the obvious signs of impending earth changes that we were not supposed to know about

Figure 4-5: The Sun erupts on Hathor. In this picture, and in many others, Hathor is depicted with Cow Ears. Hathor is likely the representation of Mother Earth (Modified from an image on the ceiling of the Temple of Hathor, Dendera, Drawing by William Gaspar.

The World Tree 101

until the End Days? In the Egyptian atlases, we also found examples of an erupting Sun on the cow-eared Hathor, the Harlot, who may have represented Mother Earth teasing Father Sky for a steamy rendezvous. Thus, the growing ears do concern us.

In the story of Marsyas, this is how the tale develops, according to H. A. Guerber:

> "A young shepherd, lying in the cool grass one summer afternoon, became aware of a distant sound of music... These weird, delightful tones were produced by Minerva, who, seated by the banks of a small stream, was trying her skill on the flute.
> —H. A. Guerber, The Myths of Greece and Rome

As we mentioned earlier, we have discovered another example of a goddess, this time from Greco-ROMAN times, whose name contains the VA: Min-er-VA! The word *miner*, a person who works in the mines, is the same in both Spanish and English, although with a different pronunciation. **The MALE FERTILITY GOD with erect Phallus in Egypt is 'MIN'!** Was the Union of MIN with VA one of the ways the Ancients hid the secrets in the name of MIN-er-VA? Well, knowing that the VA stands for the Sacred Feminine, we are not surprised about the Egyptian male god MIN being the MINER of those treasures. As we are searching for the MIN word in major star constellations, we find that the famous Belt of Orion consists of Three Stars. Generally, in metaphysical circles theses stars are accepted as the Three Kings or the Magi from the East. Orion is certainly number one on our list along with Hercules to fulfill the requirements of the Giant Hero of Astronomy and Mythology who symbolically for the war against the Titans and lost.

The Three Stars of the Orion Belt in today's Astronomy are called by their old Arabic name as 1, MIN-Ta-Ka ('Belt'), 2, Alnilam ('the String of Pearls') and 3, Alnitak ('the Girdle'). We know from the Egyptian Hieroglyphic symbols that the biliteral ('two letter') word 'TA' is shown as the 'Belt with Three stars' and the 'KA' syllable is shown as 'Two Upheld Arms' that Orion definitely has. Thus, the question becomes why we have the 'MIN' syllable that denotes the Fertility God with an

Erect Phallus! Is Orion the Phallic God MIN? Is he the ROD that we read about in the Bible where the Hebrew, who reads backward - read NIM instead of MIN, thus we know him as NIM-ROD?

The pre-Christian Sumerian Magyars also had a Great Hunter, just as Orion or Nimrod in the Legend of the White Stag written by Kate Seredy. The hunter's name was MénRót that is similar to Nimrod. The MÉN word in Magyar means HORSE or even Stud, of which we are not surprised, since the Horse head Nebula is at the Belt of Orion. The magyar 'Rot' word means 'Red', similar to the German Róth ('red'). Why we are writing about this, because the Magyar word 'Ménes' means 'A Herd of Horses', and that was the first version of the name of the famous Egyptian ruler MIN, his name was MENES. Then we seen MEN and the third version was MIN. A number of words with different vowels came to mind, such as 'MaN', MoNey, MoNkey, MooN, MiNt, all of them some ways related to the initial concept. This MIN word is what the Egyptians, Arabs, Latinos and others use for the Minister and Minerva. MIN is the best Egyptian version that describes the Hero with a phallic symbol. It is not a linguistic mystery that the English Rod (staff/rod) word, the Magyar Rúd ('rod') the German Róth (red) the Spanish Rojo (Red) and a number of others relate to either a RED ROD or a Red Giant. The star constellation Orion is the Lower Brother to the South, below the Bull and by cosmic symbolism tied to the Magnetic North that has a number of connotations included. Thus, we find the Core of the Earth the Rock, the Keystone that is the body of Hercules, the Iron Rod, the Hawk, the Red Color, the Horse and the Hunter Orion, NimRod, Min with the Phallic symbol ALL tied to the Hercules, the Giant who is the Red Hero of the Magnetic North. The Red Hero will overtake and make love to the Polar North, the Princess who is dressed up in white wedding dress as a Bride should be. The Cosmic Marriage Ceremony of the Red Hero and the White Princess has the Two Kingdom of Egypt unified. That is why the Hawk headed Horus 'Heru' ('Hero') is wearing the Red and White Hat of Unification! We know that a lot of different sounding ideas are tied to this one concept, but the more we read about it, the easier it will get. It is also interesting to note that the capital city of Mongolia is called 'Ulan Baator' that means 'Red Hero'. Thus, the Red headed westerly man does not mean ethnic designation in mythology, but it hides the secret of the Four Directional Colors of our rediscovered Cosmic Model.

The earliest and most famous Egyptian Male God's name, MIN remained in a number of aspects of our lives where the male authority dominates. The MIN-isters of political offices, the MIN-isters of the Churches, the coin and money makers of the MINts and the MIN-arets of the Arabs all stayed around for thousands of years to remind us the concept of the power of MIN (MeN & WoMeN). But to have the Cosmic Union idea in our languages, we invented MIN-er-VA, who would be the female Prophetess in Rome.

Minerva is not the only proto-typical female seer. Commonly, the females with lose morals had to be around for the myth makers of the male power to exert itself. These secretive female goddesses are present in every part of this world. The Russians and the Magyars call the hooker or street girl Kur-VA. This is the original Latin word from which curva or curve is derived and that mirrors the concept of the Hooker in English. The sacred VA is coming closer to revealing all her secrets. We will not rest until we find all aspects and all parts of this deep secret of the ancients.

As we learn more about the story of Marsyas in H. A. Guerber's work, we find out that the beautiful Minerva looked into the limpid waters of a small stream—as Snow White's mother would look into the mirror—to find her face less than desirable. Minerva dropped her flute into the water.

Before we peer into the Greco-Roman tale, let us think about another story that many more of us grew up with. Naturally, most of us heard or read about the evil stepmother of Snow White who, similar to the beautiful Greek goddess, Athena or her Roman equivalent Minerva, looked into the mirror to see herself as an ugly witch. Possibly, even more than an ugly witch, she saw a face of the devil looking back at her from the mirror.

We assume that this is a cosmological reminder of what happens when the Old Mother Earth reaches the half cycle switch of the Galactic Heart and, in this rare alignment, the Sun erupts. A red, poisoned apple is her last attempt to prevent the New Princess from taking power. The Red Apple is the ancient synonym for the geo-dynamo of the Earth, with its core being Core of the Earth.

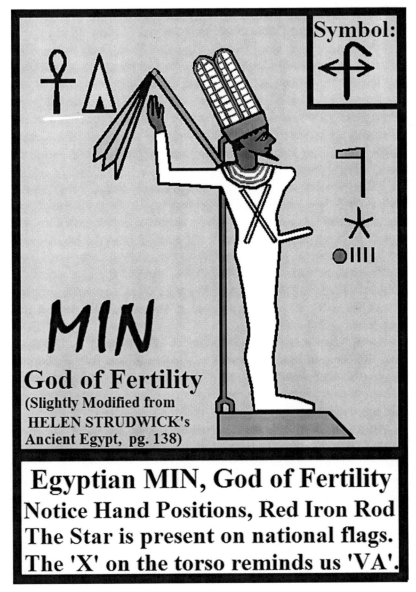

Figure 4-6: The earliest Egyptian Male Fertility god MIN. His erect Phallus represents the Male Eruptive Power of Nature during the Earth changes every 12,000 years. (Drawing by Wm. Gaspar)

The World Tree 105

The Core of the Apple and the Philosopher's Stone, as much as they sound like two different concepts, both hide the ancient secret of an active Core of the Earth's geo-dynamo. Without that core, we would not have the ability to maintain life on Earth, for a number of reasons that the astrophysicists provide us with. Without that magical stone / core, we would not have the Gaspar de CORIOLES Forces protecting the Earth from harmful UV radiations. Also, we would not have inner heat production on Earth, and we could freeze to death.

Let's return to the story of Marsyas. In numerous tales from the past, we find references to a Keystone, the Stone of the Philosopher, and the Three Stones that the Maya household keeps in the kitchen.

Some of the ancient cosmic legends survived in a childishly entertaining non threatening matter to keep the idea of the half way switch as a mirror and the passing on the Kingdom to the next young Queen. Ugly Mother Earth is reborn into a beautiful young Princess.

We have seen countless of stories from Slavic, Magyar, Scandinavian, African, Native American Indian, Arabian and classical Greek sources, to conclude that almost all of these tales present elements of the

Figure 4-7: Scandinavian / Viking Rock Carvings / Germanic Coins paralleling the Egyptian Cosmic Message in Europe several thousands of years ago. (Drawing by Wm. Gaspar)

Figure 4-8: The ugly witch Queen step-mother of Snow White (Drawing by William Gaspar, modified from Disney's beloved tales.)

universal mono-myth. The famous classical Greek mythologies are likely the best known and read stories available today. Thus, we return to our Flute flinging Greek hero Marsyas to see what we can learn from that tale.

Marsyas, the young shepherd, saw the flute floating in the stream, and quickly retrieved it. He started playing magical flute music on it and claimed that he could play better music than the god Apollo, and challenged the old deity to a musical duel.

Intending to punish him for his presumption, Apollo, accompanied by the nine muses, the patronesses of poetry and music, appeared before the musician and challenged him to make good his boastful words . . . Apollo joined the harmonious accents of his godlike voice to the tones of his instrument, causing all present, and the very Muses too, to hail him as conqueror . . . According to a previous arrangement—that the victor should have the privilege of flaying his opponent alive—Apollo bound Marsyas to a tree and slew him

The World Tree

cruelly. When the mountain nymphs heard of their favorite's sad death, they began to weep and shed such torrents of tears that they formed a river, called Marsyas, in memory of the sweet musician.
—H. A. Guerber, The Myths of Greece and Rome

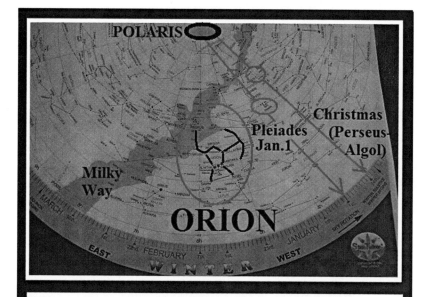

Figure 4-9: Orion in the winter. His name in the Bible is Nimrod. If we subtract the Rod from his name and read it backwards – it reads MIN. Likely a coincidence. This astronomical arrangement of the sky is based on looking toward the North Star, Polaris (Bear / Anubis) and it provides us with important dates to events of mythology! This is also how the Egyptian Dendera Zodiac is arranged. Drawing by Wm. Gaspar)

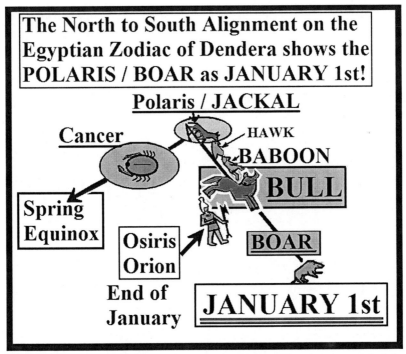

Figure 4-10: The astronomy of Christmas to Easter. This drawing is slightly modified from the outline of the Egyptian Dendera Zodiac. The Baboon stands for the Perseus/Mithra that lines up with CHRISTMAS. The shoulder blade of the Bull, PLEIADES, is lined up with January 1st. Orion aligns around the third week of January, at the time of the Chinese New Year. (Drawing by Wm. Gaspar)

Our astronomic conclusion of this tale leads us to the bank of a stream—as we so often read about in the Torah and the Old Testament—and it brings the star, Rigel, into view. Rigel is the FOOT of ORION, and starts the longest star constellation in the sky, named the RIVER. In the above story, the tears of the mountain nymphs formed a river. The name of the shepherd, Marsyas, sounds similar to marsh land, and it might take us back to the idea of the eruption of underground fountains as one of the initial events before the axis shift would happen. Let us see, then, the astronomy of Orion in the sky

The World Tree 109

when we are looking toward the North Star, Polaris, and the Bear for the Nordics, and Anubis, the Jackal, for the Egyptians.

The Wild BOAR of the Egyptians, and probably the Hindu Boar A-VA-Tar is ALIGNING on January 1! Thus, maybe, Pope Silvestre, who mysteriously changed the date of the Gregorian calendar's New Year to JANUARY first, was aware of more enigmas of cosmology then we can imagine. Was the Pope then and may be even now an initiate of the Cosmic Mysteries? Frequently, as we encounter these enigmatic dates and amazing coincidences of cosmology we wonder how much truly the Holy See knows. Following the excellent books of Dan Brown, increasingly more people subscribe to the concept that the Popes themselves may have been initiated into their own secret societies, not on the level of the Knights of Columbus, but at a much higher degree of Cosmology. The unusual assumption may have some merit as we learn more about the interests and the secrets of the Vatican. We subscribe to the credo that anybody, including the Popes of the Holy See, was keeping the cosmic knowledge alive throughout the ages for altruistic and noble purposes for the future of mankind.

Thus, the cosmology may proceed this way:

On Christmas Eve, the first amazing light appeared in the night sky as a warning sign for everybody on our globe. This likely happened on December 25th, 2012 or the year 2013 and began another predetermined awful, but required metamorphosis in the history of humanity and Mother Earth. This event likely signaled the birthing pains to the maturation of the Sacred Child of the Sun, as the ancient Egyptians would claim. Then, eight days later—at the time of circumcision of the Hebrew boys— the Egyptian and amorous Hindu Boar gave its first knock to Mother Earth. The day was December 31st, in the year of 2012 or 2013. Less than a month later, once the shield of Orion failed and the eruption of the fiery Sun broke through, the ancients also experienced a watery eruption of the underground fountains. The Core inside Mother Earth moved just a tiny bit, enough to apply pressure on the upper crust of the Third Rock from the Sun.

The future calendar showed the date January 26, 2013 or likely the year 2014. This was also the beginning part of the intense heat as the Chariot of Fire / the Chariot of the Sun departed in the sky on a deadly mission. The Charioteer star constellation, whose body contains the star **Capella with its** goat-like DEVILNESS, provided us with another clue about the timing and the origin of burning in the eternal fires of Hell. In the next few years, in the spring the Axis was ready to shift from the Magnetic North to the Polar North and therefore the Lake of Fire of the Fifth Hour of the Book of the Dead caused the land and the underwater Volcanoes to erupt and color the oceans red and cover the sky with a dark cloud. The summer time brought us the wormwood of the Scorpion and then a little more than three and a half years later the sacred life giving Light appeared again and the axis of the Eart moved again. Did it happen in a few years as we assume from 2013 – 2017 or did it take more than a few years? According to mythology, the first Big Hit from the Sun happened in the Fourth Hour and then 9 - 10 years later, at the end of the 12 years the Big Flood happened. This time frame coincides with the words of the Torah and the Bible, when Noah was given a 10 year warning to build his Ark.

This sequence of possible events is what most mythology marks with endless stories about warring gods and clever astronomical animals. The argument about certain specific years is up for grab. Our job is to provide the seekers of Cosmology with a working model. Anybody, who possesses a deeper understanding into the specifics of the earth changes, is certainly welcome to improve on our theory and correct our well meaning loose assumptions with the correct scientific facts.

For some, our writings may appear as pure conjecture or insane guessing, but our work is based on more than two decades of research into the scientific and cosmological studies of the parts of the Universal Mono-myth. Therefore, these dates around Christmas, New Year, and Chinese New Year are very prominent time periods in the ancient sky when, about 11,680 years ago, our Mother Earth went through some challenging times when the climate and the cosmic climax of the amorous Father Sky (**PAD**) / Sun King (**RÉ**) made love to our female Mother Earth. The **SACRED COSMIC MARRIAGE**, as the wise Ancients and a recent book by William Gaspar would name this time, borrowing the term from the Egyptians and the Babylonians.

The Sacred Cosmic Marriage concept can be presented in a number of different ways, not always necessarily in the form of a wedding. Interestingly, in a parallel thought we can detect the idea of the harpist

The World Tree

tied to the movement of the millstone (Core of the Earth) in the next quote from the Scripture. Then in the following quote we will hear about the harlot. This is how the excellent book of Revelation talks about the harpists.

The sound of harpists, musicians, flutists, and trumpeters shall not be heard in you anymore. No craftsman of any craft shall be found in you anymore, and the sound of a millstone shall not be heard in you anymore."
—Bible, Revelation, 18:22

First in the above verse, we hear about the Harp, then the Craftsman, and following that, the Millstone that, in the Finish Epic of the Kalevala, means earth changes and a possible axis shift. Then a paragraph later, we hear about a harlot.

2: For true and righteous are His judgments, because He has judged the great harlot who corrupted the earth with her fornications; and He has avenged on her the blood of His servants shed by her.
4: And again the twenty-four elders and the four living creatures fell down and worshiped God who sat on the THRONE . . .
5: Then a voice came from the THRONE,
6:[T]he voice of a great multitude, as the sound of many waters and as the sound of mighty thundering,
7: Let us be glad and rejoice and give Him glory, for the marriage of the Lamb has come, and His wife has made herself ready.
15: Now out of His mouth goes a sharp sword, that with it He should strike the nations. And He Himself will rule them with a rod of iron. He Himself treads the winepress of the fierceness and wrath of Almighty God.
—Bible, Revelation, 19:2–15,

On the Dendera Zodiac, with its Polar North orientation and centering, the Aries / Ram star constellation shows up in December, right before the Bull of New Year, thus the **MARRIAGE CEREMONY of the LAMB** may refer to the time period starting with December and ending at the end of March. Anyway, this Dendera Zodiac provides us with an incredible accuracy of comparing different mythologies from different continents to point to the same time period of the year almost to the accuracy of the day to day basis.

One of the interesting enigmas in different religions is the Black Virgin. We think, since **Venus is in the west only once every 121.5 years behind the Sun in its center,** it may hide the riddle of the **ancient secret for the BLACK VIRGIN or the Black Widow. The Year Zero for the Venus could be 2008 and the fourth and fifth year of burning would then point to the years of 2012 – 2013.** Then we discovered another VA in the vul-VA of Rocamadour:

Roland's sword, Durendal, rightly reposes in the vulva-cleft of Rocamadour and brings fertility to brides . . . The Black Virgins are often associated with esoteric teaching and schools of initiation.
—Ean Begg, *The Cult of The Black Virgin*

We should start asking ourselves if this OPERA of the heavens is not closely related to the harpies and the harp players of the classical Greco-Roman mythologies. The angels in the book of Revelation are also sounding off their Harps to warn us of danger. Since the Harp is the only musical instrument in the sky, and it is associated with the Magnetic North, it tells us that the Music here may be not a good thing. Not only can one observe the Two Sisters on the temple walls of Angkor Wat in Asia, but one must take specific notice of the fact that they have Cobras hanging off these beauties' heads. There are also other little cosmological markers.

In our theory, we think that the ending of '**RA**' was placed in the names of the Roman / Latin serpents to provide us with a clue that the Snake is representing the energy of the Sun, as a Cobra or a Rattlesnake ready to strike the Sun displayed that same character during those half precessional switches. Names such as RA-ttlesnake, Hyd-RA, Cob-RA, Vipe-RA all contained the '**RA', just as the OPE-RA contains the RA ending** to keep the knowledge of cosmology.

The World Tree 113

Now a great sign appeared in heaven: a woman clothed with the sun, with the moon under her feet, and on her head a garland of twelve stars.

2 Then being with child, she cried out in labor and in pain to give birth,

-Bible, book of Revelation 12:1-2,

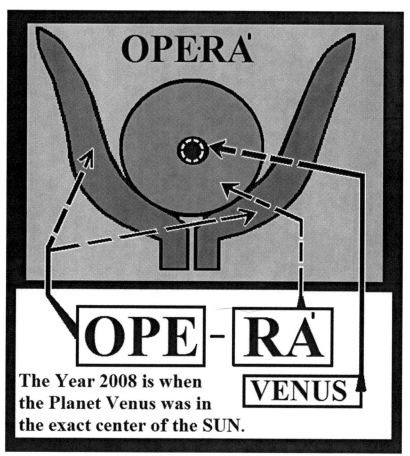

Figure 4-11: The Egyptian Opera hieroglyphic sign is hiding the same cosmological concept as the Black Virgin, the planet Venus in the west in the Center of the Sun! Today's date would be June 9th, 2008 and then the next station would be on June 5-6, 2012 (Drawing by Wm. Gaspar)

Figure 4-12: The Virgin of Guadalupe represents the Sacred Feminine. She is dressed with the rays of the Sun and she is wearing a crown on her head, and she is standing over the moon, which in our cosmic model, are the horns of the Bull / Taurus during the Venus Transit. Certainly, the west designation of Venus can be understood in different ways (Photo by Eva Gaspar)

The World Tree   115

   Very clearly we are being told that the woman is a celestial being and not a flesh and blood human. The birth that she gives also is to a celestial being. When this happens, we know the End Days are near. This Christian judgment has its cosmological roots in the teachings of the Egyptian Astronomy Priests of Rá.
   Thus, we examine again the JUDGMENT scene of the Egyptians to find that there are TWO SISTERS or goddesses who are standing behind the Pharaoh whose headdress shows the Unification, and who sits on a THRONE with a RED BRICK under it. We already know that the Red Brick under the Throne represents the Magnetic Axis shift that had to happen in relations to the ending of the Two Venus Transits. That is why we see the Two Sisters behind the Pharaoh, who is mummified and cannot move. The parallel arms of the Pharaoh also mean that the Magnetic North Axis went parallel with the Polar North.

Figure 4.13: In an Egyptian Judgment Scene, the Two Sisters stand behind the Mummified Pharaoh. These represent the Two Pillars of the Earth and also relate to the Venus Transits in front of the Sun and the time of the Judgment. (Drawing by Wm. Gaspar)

The 'GATE', the PORTAL to Heaven in Uxmal. It is between TWO SPIRALS, with 2 top Serpents- the Door make the Letter 'T'. Snakes: 4 & 7 waves. The sitting Man to the right has Paralell ARMs tied together. The Mosaic Tiles depict: 4 x 5 = 20 parts inside. Example of exact Mayan Cosmology.

Figure 4-14: *Mayan cosmology from Uxmal, The GATE. It is also present in the Mayan Cosmology and it seems to represent the Male Force penetrating Mother Earth at the end of the Portal of 2004 – 2012 (Drawing and photo courtesy of* Wm. *Gaspar)*

The Closing Shut of the Cosmic Door of Heaven is displayed in the book of Revelation of the Bible in the following manner:

And I will give power to my two witnesses, these are the two olive trees and the two lamps stand Standing before the God of the earth. And if anyone wants to harm them, fire proceeds from their mouth,

these have power to shut heaven . . .
—Bible, Revelation, 11:3-6

So, are the Two Olive Trees the Two Northern Axes of our Earth's heat producing geo-dynamo? It is a scientific fact that the Earth's Polar North is currently tilting away from the Magnetic North about by 23.4 degrees.

Naturally, the 23.4 degree tilt represents one third, and the difference remains an evil number that is 666, or the 66.6 degrees count two-thirds from the Horizontal plain.

Let us quote from the Epic of Gilgamesh as regards his connection with one third and two thirds:

Supreme over other kings, lordly in appearance, he is the hero, born of URUK, the goring wild bull . . . It was he who opened the mountain passes, who dug wells on the flank of the mountain it was he who crossed the ocean, the vast seas, to the rising sun who explored the world regions, seeking life. It was he who reached by his own sheer strength UTANAPISHTIM, the faraway, who restored the sanctuaries (or cities) that the flood had destroyed for teeming mankind. Who can compare with him in kingliness? Who can say like GILGAMESH: I am King! Whose name, from the day of his birth, was called GILGAMESH? **Two thirds of him is God, One third of him is Human."**
—Maureen Gallery Kovacs, Epic of Gilgamesh

We learn from this quote so many things that are associated with our theory, but it is important to notice the connection of the last sentence of the quote, where Gilgamesh is two-thirds god and one-third human. We definitely now know that the number 666 in GEMATRIA it is associated with the Messiah. In our next chapter, we will introduce the exact enigmatic cosmological secret of Jonah's WHALE, since Christ was teaching about the Fishermen concept and the Old Testament talks about a Great Whale. What can be the synonym for a Great Whale in this already complicated Mono-myth? We will answer that deeply held secret that our readers have never heard before from any writers of ancient mysteries.

18 Here is wisdom. Let him who has understanding calculate the number of the beast, for it is the number of a man: His number is 666.
                                    –Bible, book of Revelation 13:18,

If we look at the 90 degree angle hold of the Solar Plane on the Magnetic North and subtract the current angle tilt of the Polar North, which is 23.4 degrees, then we will arrive to 66.6 degrees. That 666 is likely the sum number we can calculate from the superb knowledge of the Freemasonic SQUARE (90°) and the opening angle of the COMPASS (23.4°). We hope that more and more members of Secret Societies, such as all degrees of the Freemasons, Knight Templars, Shiners, and the marvelous 32 degrees of Scottish Rites, along with the members of the Knights of Columbus will pick up this book and other books like this to learn the real astronomical secrets of their trade.

Average folks think that secrets are abound in these organizations just for the asking, but it is not so, only by a studious approach to study science, astronomy, art and the languages under the guidance of the right mentor will provide the seeker with some understandings of the hidden. As matter of fact, we believe that nothing is hidden only the limitations and the blinded imaginations of the seekers will stop them to further themselves in the knowledge. There is no conspiracy to hide any knowledge. Only we ourselves conspire to accept the dogmas of varied teachings that narrow our understandings of the Sacred. If one believes in an Almighty Universal Spirit God, then the spelling of the name of their 'local' gods should not stop them to look across the fence to realize that we humans fenced up the landscape, but the Almighty known under the varied names of God, Gott, Thor, Isten, Vishnu, El Señor, Dios, Jehovah, Yahweh, Allah, Kukulcan, or Wakan Tanka created both parcels on either side of the fences. The dogmatic behavior in a one 'small local god' who only serves me, but hates everybody else, will not provide the advanced cosmological knowledge that can only come from the Universal Almighty! Let us look at each other for the millions of similarities we share as humans, rather than limit our understanding by the minute differences that we created during our local customs. Nature will provide us with plenty of fireworks when the appointed time arrives, thus no need for us to carelessly allow the nuclear weapons to serve the unexplained needs of local deities.

Before we conclude this chapter, we would like to share the story about the Long Ears of King Midas, as we promised earlier:

Apollo found himself engaged in another musical contest with Pan, King Midas' favorite flute-player. Upon this occasion Midas himself retained the privilege of awarding the prize, and, blinded by partiality, gave it to Pan, in spite of the marked inferiority of his playing. Apollo was so incensed by this injustice that he determined to show his opinion of the dishonest judge by causing generous sized ASS's EARS to grow on either side of his head.
—H. A. Guerber, *The Myths of Greece and Rome*

We think that all these ancient tales talk about natural earth changes and sudden shift!

In a few years or decades, we might experience it ourselves.

Since the ASS / Donkey stars are part of the CANCER constellation, we need to look at the Cancer constellation in more details. Within the Egyptian Zodiac of Dendera, Cancer corresponds closer to the month of March, but when one looks at the Hebrew calendar, it seems to show the month of TAMMUZ. Actually, the **Hebrew holiday TAMMUZ 17**, 5769 (corresponding to July 9 in 2009) is a prominent day in the history of the Jews, marking the date when, after forty days, Moshe (Moses) came down from Mount Zion. On the regular Sun Zodiac we find Tammuz 17 in July, but transposed on the Zodiac of Dendera the date would be around the start of the spring season. Tammuz 17 is a very sacred date for the Children of Israel, because that was the day when the breach of the walls of Jerusalem happened. This same day in the past, Moses broke the Twin Tablets of the Ten Commandments. Is the secret cosmological teachings of the Hebrew Moses also point to a disaster that may have happened around the Spring Equinox? It most certainly appears that way. All we can do is to bow to the greatness of past prophets, seers, shamans and medicine people who advocated an advanced knowledge of Cosmology to save portions of the human race in that insane period of the turmoil of Mother Earth and Father Sky. No blind hatred, no evil ethnic genocide, no religious persecution should be the lot and vicious reward of those, who kept mysterious cosmic stories of human sacrifices to record the sufferings of all humanity when Mother Earth cried out with a loud shriek to give birth to the Egyptian Son of the Sun, SA-RA!

CHAPTER 5

The Chariot, the Ocean and the Sacred Twins

Tender Gemini in strict embrace
Stand clos'd and smiling in each other's Face . . .
—Manilius

In this chapter we will talk about a number of things, but the main take home message is that the Warming Up of the Earth in the past began with the Chariot of the Sun going awry, and the hottest period ensued when the Fire TWINS showed up on the Ecliptic.

One of the most amazing similarities between seemingly unrelated cultures is their twin stories. We need to emphasize that these tales are not identical with the Two Brothers, where one brother kills the other as we can detect in the tale of Cain and Abel of the Bible. Those are mythological brothers, who stand for Perseus and Orion, but they are not Twins. Then there are the occasional Two Sisters, or most likely the Three Gorgons, often presented in mythology and religion as the Harlot, the Virgin and the Widow. Of the Three Gorgon Sisters, two were immortal, but the third one could die. This, we will clearly demonstrate in our final chapter by our secret formula.

Then, finally we are aware of the SACRED FIRE TWINS of mythology. Usually, these twins are both boys, although we find the Brothers and Sisters of mythology, such as the boy pharaoh who marries his sister.

For our purposes the Greek Twin Boys, Pollux and Castor are the best known examples in our western mythology and astronomy. Now, the male and female twin stories of the Egyptian pharaohs are usually morally disturbing thoughts, especially when these blood sisters will marry their blood brothers. This scenario might make a specialist in genetics cringe, but the ancient bards and myth makers employed this distasteful Cosmic Love so we would cringe and cringe again. Then, when we cringed enough we would remember the sick tale. Remember will do, because it so awfully sickening to imagine that if it happens in real life it is very rare. Since, love affairs can be a little bit cliché the love affair of siblings would create some disturbing memories. Human psyche is a very interesting and complex beast.

Now, after this sickening thought, we can say that besides the Vulgarity, the titles of Politicians, Church Fathers and Ministers, we can surely add to the lists of secrets the stomach turning and disturbing marriages of the pharaoh siblings to the armament of the Makers of these cosmological Secrets.

When these Astronomy Priests needed us to remember something very important then they had the pharaonic brothers and sisters to marry, or they had the Old Kings to have sex and fornicate with either Harlots or Virgins. By far, the worst and most distasteful cosmic sex scenario that the Ancients advocated as pure history was the Incest between a child and a parent. Sure, a lot of times the sex was not between humans, rather they had the tricks of having the gods turn into animals, such as Prajāpati into a Deer who had a daughter born by his first daughter by incest. One of the better known and historical sounding tales was the legend of Oedipus and his birth mother, whose likely fabricated cosmic love story strangely defined the development of the age of psychiatry. Was it history or not, we do not know. For our purposes, all of these stories hide a legend that happened to Mother Earth through the sudden and unexpected action of the 'horny' old Sun King. Not only the horns of the astronomical animals that are separated or singular, but the split hooves against the full hooves make a bit of a difference to mark the period whether the two pillars are together or united.

The Myth Makers all around the world succeeded in making non-existent wild creatures to invade our minds and heart, to feel up our

The World Tree 123

imaginations to the rim. It really did not ever matter to the light-hearted audience, it still does not. The made up magical steeds and guilty monsters, such as the Unicorn and the fiery Dragon clearly represented the disastrous axis shift. The myth makers and secret keepers only cared about carefully installing their precise Cosmic Disaster Scenario into the history books and the fairy tale magic. They did not care about the genetic outcome of those imaginary marriages of the sisters and brothers of the pharaohs, since the story was about a cosmic union and not actual people. They also did not have to explain to the young girls that it would be dangerous to ride on a one-horned Unicorn, because it was not true, thus not of any consequences.

The only thing that the Myth makers of ancient mythology were truly concerned about is how to insert part of the secret Cosmic Scenario into any great stories. The Masters only cared if the strange legend was portable throughout the ages. If the legend could become a favorite with our kids and even the adults, if it can also fulfill the requirements of religion or spirituality then the story is a success. Thus, the amazing tales of steamy COSMIC UNION has been turned into a historical Sacred Cosmic Marriages that in reality has never happened.

Reality, logic, rational or historical authenticity also did not matter to the celebrating crowd of the ancients Egyptians, or the participants of the North American Mardi Gras celebration, either. When a rowdy fiesta was established, everybody won. The myth makers propagated their strange cosmic stories in secret, the celebrating crowds threw tons of confetti in the air, walked on fresh flowers, got drunk, sang and danced and most importantly they made love to anybody they pleased that night. Therefore, everybody achieved what they were aiming for. The scrooges of myth makers propagated their stories and the crowd had incredible fun.

The most serious scenario of the cosmic disaster began on one December night almost 12,000 years ago on the future birthdays of Krishna, Mithras and Christ. This happened when the four directional Northern Cross stood straight up in the sky and the Devil star, ALGOL aligned on December 24 - 25th, a warning bright light appeared in the night sky. It was seen from all parts of the globe and it initiated a swift course of events that devastated the Earth. This infamous star named

Algol from the Northern brother Perseus/Mithras above the Bull was known to the Hebrews as '**Rosh ha Sãtãn**', the Head of Satan. The Arabs called it the Demon's Head and the Greeks, who rediscovered the Precession of the Equinoxes, connected it to the ugly looking cut off Head of Medusa. Medusa, the only mortal sister of the Three Gorgons was killed by Perseus. The story could fill another book. The infamous Algol star in the shamanic hat of Perseus was named by the Chinese as **Tseih She**, the PILED UP CORPSES.

What happened on that awful night when heroes, saviors and redeemers were supposed to born? What occurred on that faithful night that was so carefully marked by the Egyptians, Hindus, Babylonians, Jews, Greeks and Romans? What did the birth of that first bright light in the sky caused that piled up corpses somewhere on Earth for everybody to remember? Was it a small eruption of the Sun toward the Earth, which may have caused tsunamis and piled up the corpses of the people from the shores? Maybe, it was a huge fireball from a large meteor that disturbed the harmony of the waves? The ancient myths and religious writings give varying clues to the nature of that old disaster, but our religions kept that late December date alive by the numerous saviors whose birthdays we so lovingly celebrate. The festivities certainly should go on, at least until the end of December of 2012 or 2013, when the mythological gods would likely pile up a few more million corpses for us to talk about.

Innumerable spiritual leaders, artists and priests from past generations made us remember the occasion when Algol aligned. They propagated the story with the pointy Phrygian Hat of the Siberian shamans, Keltic druids, Santa Claus, Arab Imams, Robin Hood, Peter Pan and thousands of others.

After this cosmological alignment in the past, the Magic Bull arrived in the sky that was equated with the Seven Sister stars of Pleiades. The Seven Sisters were equated with the Shoulder Blade of the Divine Bull that was targeted by so many heroes, toreadors and matadors. The secret date of the Pleiades is December 31st. The first and the oldest Bull related story comes from Egypt and the knowledge of that had to be passed down to the great Roman Empire for Pope Silvestre to change the New Year of the Gregorian calendar to that date. In Egypt the Male Fertility God, MIN had a White Bull for a symbol. Then, we

The World Tree 125

are not surprised that this White Bull shows up in Greece as the White Bull favored by king MIN-os. Even the name MIN, the name of the rulers had changed little, and to the list we can also add the Roman Goddess MIN-erva. Have we received the well rehearsed cosmological tales from all the outposts of the Egyptians, but with little different names and slightly differing ethnic stories? Yes, very likely!

We know that since the successful killing of the Bull Headed Babylonian Humbaba, protector of the sacred Cedar Tree of Lebanon,

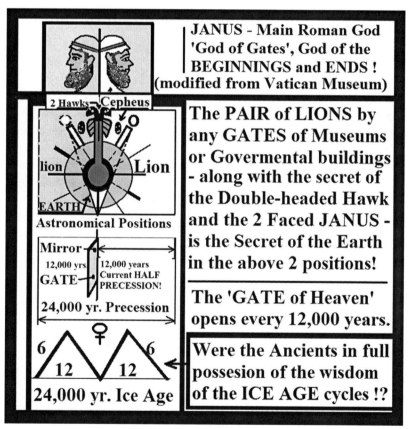

Figure 5-1: *The Sacred westerly view from the Heart of the Galaxy shows the 'Twin' Lions, the Twin Faces of Janus, and the Twins of the overheating Mother Earth around the Latin Holiday of Love on February 14 – Valentine's Day (Drawing by Wm. Gaspar)*

the Bull, the Hero and the felling of the cosmological Tree is closely connected. That sacred Axis Mundi of Mother Earth is only targeted every 6,000 to 12,000 years by these relentless heroes of mythology, thus we have plenty of good years to celebrate the fact that this year our religious trees are still standing straight.

On the astronomical charts, after passing the Shoulder Blade of the Taurus Bull, we encounter the War Chariot of the Sun and Fire. Following this fiery ride, the industrious Gemini Fire Twins begin to twirl their fire stick to heat up the inside Core of Mother Earth. Then the Axis shift of the star constellation of the Cancer and Leo will complete the cosmic ceremony of the 'Prince taking over the Kingdom', the Sacred Cosmic Marriage or the astronomical reason for the **'Marriage of the LAMB.'**

The main star on the Ecliptic, Regulus from the Lion star constellation defines the Horizon in this model. It is at this point where we mark the sinking of the Lion from three o'clock to four o'clock during the Axis shift. This pair of Lion that we have on this west to east view model is what the leaders of our empires commemorated by keeping Two Lions by the entrances of Museums and Ministerial buildings.

Other famous and obvious places where one can observe the Lions symbolizing a nation are on flags and heraldic shields of the nobles of Europe. If one remembers the coffin of Princess Diana, it was wrapped in a yellow linen displaying Three Red Lions stacked above each other. The Heraldic depictions of the British Royalty, the Prince of Wales, the House of the Lords, the Royal Arms of England, the Bishop of Norwich, the Hapsburgs of the Astro-Hungarian Empire, the kings of Spain and Italy all displayed a Lion, a Hawk even with two head in some form of knowledge and understanding of Cosmology. These secrets of animal astronomy were proudly displayed at events sponsored by nobles, priesthood and chivalry without the fear that the commoners, who prayed to a human god would object to the seemingly pagan depictions. The Fire arrived in the Fourth Year and the Lion roared in the Fifth year.

We begin to acknowledge that the first signs of troubles in that awful fourth year amongst the last 12 years lasted from the month of December to about March and it consisted of three very brutal

The World Tree

months. The memory of these dreaded events from the past is what the Egyptians, Babylonians and the Jews celebrated in their ancient mythology as the Sacred Cosmic Marriage! These very memorable and eventful three months of religious writings cover some serious events.

23: By faith Moses, when he was born, was hidden THREE MONTHS by his parents, because they saw he was a beautiful child; and they were not afraid of the king's command.
24: By faith Moses, when he became of age, refused to be called the son of Pharaoh's daughter . . .
31: By faith the HARLOT Rahab did not perish with those who did not believe . . .
–Bible, Hebrews 11:23, 24, 31

These are cosmological teachings that the Hindus, Egyptians, Nordics, Natives, Greeks, Romans, Jews and Christians left to us. The knowledge of the birth of the Hindu Krishna, Lord Skanda, Mithras

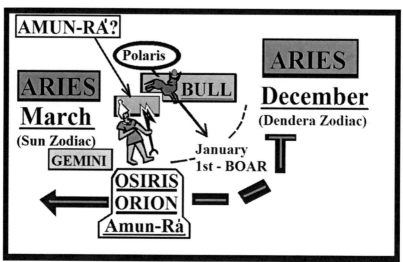

Figure 5-2: The Two Aries / Two Lambs / Rams are on either side of Orion. The first one ends in December on the Dendera Zodiac, and the second one begins a season later in March. In the Middle of it are the Fire Twins of the Gemini star constellation (Drawing by Wm. Gaspar.)

and the Christ is a great wisdom, if one knows the exact symbolic meanings to the important cosmological parts of the puzzle. Nobody should be afraid to read the wisdom of other religions, since it is all ONE with common roots and common cosmological wisdom behind the historical meanings!

The time of the Birth of the Sacred Son of the Sun, along with the Sacred Cosmic Marriage in some form or another was joyfully celebrated in most cultures. The secrets of the sad reminder of the earth changes were carefully hidden from the uninitiated, thus the festivities and the strange magical happy tales kept going on. Expectedly, this was another amazing trick of the Ancients; - the astronomy priests learned early on that they can pass on any falsities or hidden truths to the masses, if the final outcome was a nice joyful celebration. The pious Egyptian crowd of old times would go out to the streets to celebrate the Sacred Cosmic Marriage of the gods displayed by the living Pharaoh and his sister wife. The boat of the Harlot, Hathor the goddess of sensuality was dragged up on the river Nile against the currents for two weeks by a cheering crowd. The belly dancers and other scantily dressed female participants of these unusual religious celebrations could drive the intoxicated males just as wild, as the beautiful harlot, Mother Earth could tease the old Sun King into an eruption.

The dressing, the religious regalia, the songs and prayers, and even the very peculiar mannerism were very infectious to come up with the right act to please the gods with an unusual taste. This was the winning formula for everybody, why change it? The only time this affair became obviously one sided, when the unexpected relentless and brutal changes arrived from a reversing Nature. Unfortunately, it was too late to reconsider; the Pharaoh and his loyal astronomy priests were safely hidden underground with all the available food storages in their possession.

In Egypt, it was Osiris the Wasir who carried the title of the Lamb of God or the Ram of the Sun, Amun-Rá. This same Osiris / Orion / Amun-Rá represented the spiral flat Horns of the Sacred Ram that reached from December to March on the astronomical chart. This was the time when the Old Mother Earth gave sacred birth to the Son of the Sun, before she needed to metamorphose into the New Age. She was symbolically that wrinkled up mother of the Old Testament that could get pregnant even at an old age of ninety nine. At this

The World Tree 129

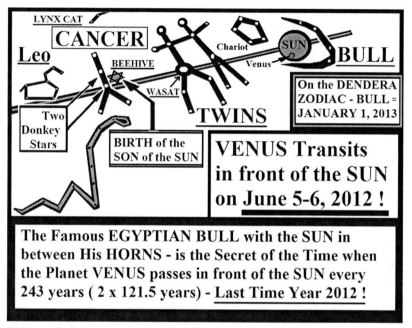

Figure 5-3: The Twins, Pollux and Castor of Astronomy on the Ecliptic. Worthy to notice the name of the Star 'WASAT', which means 'the middle part' in Arabic Astronomy, but also a reminder of the Iron Rod that is named WAS or VAS in Egyptian. Is this a reminder for reaching the Middle of the Iron Rod of the Earth? (Drawing by Wm. Gaspar)

rejuvenating time the Old Serpent shred its skin and a new Serpent became in control of the Earth.

    15 Then God said to Abraham, "As for Sarai your wife, you shall not call her name Sarai, but Sarah shall be her name.
    16 "And I will bless her and also give you a son by her, then I will bless her, and she shall be a mother of nations; kings of peoples shall be from her."
    17 Then Abraham fell on his face and laughed, and said in his heart, "Shall a child be born to a man who is one hundred years old? And shall Sarah, who is ninety years old, bear a child?"
--Bible, Genesis 17:15-17

We know that historically God kept His promise and His Covenant with the Jewish people, who were singled out by many Emperors throughout the ages for complete annihilation and the more adversity happened to them the stronger they became as a nation. Also, we can readily detect the emersion of meaningful teachings of astronomy and cosmology into their historical journeys.

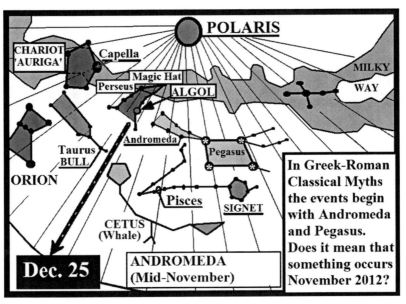

Figure 5-4: Perseus, Cepheus, Cassiopeia, Andromeda, Pegasus, Pisces and the Whale (Drawing by Wm. Gaspar)

When Tamar, Judah's daughter-in-law was widowed it was difficult for her to see ahead that she will soon have twins born to her. The name TA-mar may also hide some very important cosmological knowledge. The fact that Tamar was given a Signet Ring (Circlet of Pisces) by Judah, tells us that she might be the astronomical representative of the Princess Andromeda.

13 And it was Tamar, ...
14 So she took off her widow's garment, covered herself with a veil and wrapped herself, ... 15 When Judah saw her, he thought she was a harlot, because she had covered her face. 16 Then he turned

to her by the way, and said, "Please let me come in to you"; for he did not know that she was his daughter-in-law. So she said, "What will you give me that you may come in to me?" 17 And he said, I will send a young goat from the flock." So she said, "Will you give me a pledge till you send it?"

18 Then he said, "What pledge shall I give you?" So she said, "Your signet and cord, and your staff that is in your hand." Then he gave them to her, and went into her, and she conceived by him. ... ... 27 Now it came to pass, at the time for giving birth, that behold, twins were in her womb.

–Bible, Genesis 38:13-27,

The Signet Ring was and still is a very famous and powerful tool of mythological Kings and real Kings and Popes alike. It is the shape of the Beehive and is tied by the end of the Cord, in astronomy named Al Rischa between the two Fish of Pisces. The powerful Cord holding to Fish astronomically is straight below the Princess Andromeda. Right below that is the Whale, Cetus of astronomy that will become important later on. This struggle of the two fish pulling on the rope, maybe a reminder of the power struggle that was evolving between the positive and negative forces of the sky. The Signet and the Cord are definitely clear astronomical markers of this story of the Old Testament. The Staff will be the one that is in the hand of Osiris – Orion on the astronomical chart. Well, history aside, this ancient tale has the cosmological reminders of an older male god and the harlot. On top of everything, she is pregnant with TWINS.

We assume that these stories are the secret representations of the ancient Earth changes from Andromeda to Gemini, because the next chapter of Genesis introduces one of the most famous biblical mortals whose name is Joseph in Egypt and who has foreseen the coming of the Seven Years of Famine. Now, just as Joseph saved his brothers and the Pharaoh, he can save us today. That can be accomplished, if we take the ancient Egyptian and Hebrew tales seriously and imagine that the Seven Years of Famine can soon arrive to our generation. Certainly, if the earth changes materialize soon, as we expect it, then even the very well prepared private survivalist groups will have a difficult challenge to keep abundant food, shelter and clean water safe from the elements and marauders.

On the Egyptian Dendera Zodiac, in the next figure, we shall demonstrate that in Hindu, Egyptian, and Mayan mythology, the **FIRE TWIN BROTHERS** were shown drilling with their fire sticks, apparently to heat up the Earth's geo-dynamo after the Chariot of the Sun / Fire started getting too close to our globe in that faithful year.

Apparently, in classical Greco-Roman mythology, it was Apollo, the Sun God, who made love to Clymene, a fair nymph. She conceived a son whom she called Phaeton.

Taking the "Ph" off from the name of Phaeton, we are left with

Figure 5-5: The Fire Twin Brothers from three different continents, from the Egyptians, Mayans and the Hindus are all showing that the Earth heats up when the Fire Twin Brothers line up in the sky (Modified from John M. Jenkins, Maya Cosmogenesis 2012 by Wm. Gaspar)

The World Tree 133

AETON. This is very similar to **ATON** from Egypt, who was depicted as the **ERUPTING SUN.**

Thus, Aton would correspond to the Fourth hour of the Book of the Dead with the Sun eruption and Amun-Rá would then stand for the Fifth hour when the Axis would be shifting. Therefore, Aton was as much a Solar Deity for the Egyptians with its Sun eruption as Phaeton was for the Greeks. Although, linguistically we find even closer correlations in the Egyptian, Hebrew, Greek and Roman civilizations, where the Fire Lord of the Sun, ATON maybe found in Hebrew as ADON (='Lord') and the biblical fiery ATONE-ment, while in the Greek mythology we find ADON-IS and the Latin Nobility is DON. In the name of the Greek god Adonis we find the 'IS' ending that in the ancient Sanskrit language means God and it is found widely distributed in the world from Osir-IS and IS-IS to the Arab IS-hmael and the Hebrew IS-rael and the Magyar IS-ten (=God). Thus, it is no surprise to see a handing down of Cosmology from the Egyptians and the Hebrews, to the Greeks and the Romans, although we sense that all of those nations have been around in that region for a very long time.

In that classical Greek mythology tale, growing in fame, Phaeton (Ph-AeTON) the Son of the Fiery Sun became conceited about the status of his father as the Sun King and boasted about it in school. The kids made fun of him until one day he went home crying. Finally, he was so infuriated about it that he asked his mother to show him how to reach his father, the Sun. One sad day, the mother of the youngster, Clymene, told her son, Phaeton, how to reach his Brilliant Father. The impatient Phaeton full of the feeling the heat of the revenge, quickly went to talk to his father, Apollo / Helios the Sun King. The youngster so begged his father to one day allow him to guide the Chariot of the Sun in the sky that the Old Man Sun promised him to one day allow him to be in control of the Fast Steeds of the Sun. Expectedly, the hot headed young Phaeton was not careful enough and drove his father's Chariot too close to the Earth and completely burned it up. Everything turned to ashes. Everything! Even the green things burnt up. We are not surprised as these old stories are full of references to the fiery oblation that pleases the gods.

Besides the Egyptians, Hebrews, Greeks and Romans, do we have this war Chariot of the Sun also rooted in the Siberian Shamanism? In

Figure 5-6: The most obvious depiction of the Egyptian Ram God is AMUN or AMUN-RÁ, who is the Egyptian Ram, likely representing the time period between December 21st until March 21st (Drawing by Wm. Gaspar)

The World Tree 135

our next figure, we will be able to see from a Siberian Shamanic drum, that the idea of Santa Claus likely has its tales originated from Pagan shamanic ideas of this Cosmology and Mono-myth we are building up step by step. On the Siberian shaman drum decoration from the Sölkup tribe, the depiction of the smaller Sacred Tree is strangely shown growing out of the Sun. Thus, we assume that the Tree that was connected to the Sun was the Magnetic North Axis of the Earth, the one that the Hawk sits on, the one that is known as the biblical Iron Rod of rulership and also the one that moves the Keystone Rock inside the belly of Mother Earth. Have the smart Siberian shamans maintained an important Sun and Magnetic North related cosmological piece of information? Are they trying to show us that the sun stabilizes the magnetic north, the Rock Core of the Earth? Thus, Santa Claus' Reindeer-drawn sled is nothing else than the War Chariot of the Egyptians or the famous Chariot of biblical Ben Hur cosmologically? It is very plausible.

Thus, this international race of the swift Chariot of the Sun is the prelude for the severe heating up the geo-dynamo of the Earth. The first Big earth change events that is sudden and global, likely will happen in the Fourth Hour of the book of the Dead and it begins on December 25th 2012 or 2013. We count the Last Twelve Years beginning with the famous Venus Year of June 9th, 2008 that will be our 'ZERO' Year. Thus, the year 2009 is the First Year of the Last Twelve, then 2010 is the Second, 2011 is the Third and 2012 is the Fourth Year! That is when the First Big Global changes, apparent to everybody should be starting if our count is accurate.

The famous star constellation that occupied the dark sky on the date of December 25th is naturally Perseus, the northern brother above the Bull. His main star is Algol, the Demon Star whose name in Arabic is Al Ghül, the Mischief-Maker. The name Ghoul appears in the beautiful stories of The Arabian Nights. Only the newest versions of the Old Original Solar Religion were not prepared to handle this most important date with the necessary and obvious horrible elements of mythology. The classical Greeks were not afraid to initiate the adventures of Perseus that would eventually end in the decapitation of Medusa. Let the blood squirt where ever it can. 'Perseus et Caput Medusae' was a hero who promised the evil king Polydectes that he would bring him the head of Medusa, one of the Three Gorgon sisters, if the dishonest king would not try to marry Perseus' mother. The ugly Medusa was the only mortal sister amongst the three Gorgons. As matter of fact, for us she equals cosmologically the Black Virgin of various cults and religions, as we will demonstrate this concept in the conclusion of our book.

The young Medusa of the Greco-Roman mythologies was originally a very beautiful young girl with flowing red hair. Naturally, the god of the Sea Poseidon / Neptune fell in love with her, kissed her and

On this Shamanic Drum of the Sölkup Tribe - one can notice The WORLD TREE - then the Second smaller TREE representing the MAGNETIC NORTH. The 'LAME SHAMAN' (blacksmith) rides the Chariot of Santa Claus. Seemingly, our religious heritage is much older than we would like to believe (from the excellent book of 'Sámánok / Shamans' from Hoppál Mihály).

Figure 5-7: The World Tree and the Sun Depicted on a Siberian drum (Modified from Mihály Hoppál's book, Sámánok (Shamans), (Drawing by William Gaspar.)

The World Tree

even made love to her in the Temple of the goddess Athena / Minerva. Witnessing the sacrilegious act, the goddess Athena / Minerva became angry at Medusa and changed her hair into hissing serpents. After long adventures Perseus was able to cut her head off. The elements of the Mirror used in this and other tales are secretly representing the Shield of Orion that held back the attack of the Sun. This defense was only successful until the last week of January of that faithful year, around the Chinese New Year when the Sun eruptions broke through as the 'kids' of the famous astronomical Goat, Capella, the foster mother of Zeus.

On the Siberian drum, the Cosmic Tree was strangely shown growing out of the Sun. This giant World Tree in the center of the drum symbolized a very important celestial crossroad for the shamans. It is the place where the Milky Way Galactic Plane crosses the ecliptic, which is the path of the Sun. This is where the westerly Galactic Plain meets the easterly Solar Plain. This place now is the actual Chariot constellation in the sky. The three horizontal lines on the bottom of

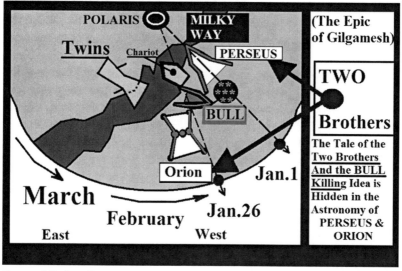

Figure 5-8: Our Cosmic Model shows that the Chariot constellation follows the Two Brothers and the Bull. After the Chariot comes the Fire Twins, then the Cancer with the Beehive. The Gemini Twins constellation sits in the middle of the Milky Way on the above graph. (Modified from Star Talker Astronomy Chart by Wm. Gaspar)

this World Tree suggest that the Galactic, Solar, and Earth lines are connected by this Magical World Tree.

The Two Sisters are the Two Pillar of the Earth and tied to the Venus Transit Pair in front of the Sun. The secret Cosmology of those two and Three Sisters will help us rediscover the enigma of the Cosmic Form that hides the totality of most of this knowledge.

Other than his deed as the Hero who severed the Head of Medusa, our savior Perseus is also known by his struggle to save the fair princess, Andromeda, who was the daughter of the Queen Cassiopeia. The old Queen was so vain that she thought she was fairer than a sea nymph, more beautiful than her own astonishing daughter. Her story is clearly resounding in the modern tales of the Ugly Witch, the mother of the princess Snow White, or even in the story of Cinderella. To punish the vain Queen Cassiopeia (whose Egyptian Throne we already know from astronomy), the princess Andromeda (andro / androgynous –mead / middle?), whose name refers to male characteristics of a female, was tied to a Rock on the coast where she was exposed to the devouring sea monster, the huge water snake. In other stories it would be the 'eunuchs' of the king who would represent the changing forces from male to female.

The star constellation Andromeda is right below the Throne of the celestial Queen, Cassiopeia and the King Cepheus, the Head who will be 'beheaded' in astronomy. The Princess Andromeda's body in the sky (November 10th –December 10th) is tied to a celestial Cubical Square Rock that is named (October 25th – November 10th) Pegasus, the larger Horse following Equuleus, the Foal (September 21st – October 1st) indicating that warnings signs of the impending changes will be around by the last week of October 2012-2013.

Even if the more serious earth changes only materialize by December 25th, 2013 – we can expect warning signs much earlier. By the markings according to the Book of the Dead, the uneasiness might even be noticed beginning in the spring of 2012 when we can begin to sense the amorous celestial gods wanting to unite.

A story cannot be told without monsters, beasts and heroes who will win the hands of the princess. Naturally, our undefeatable hero, Perseus came to the rescue in the past and slaughtered the slimy evil Sea Monster and gained the hands of Andromeda in exchange. Hopefully,

The World Tree 139

with such a name she turned out to be a beautiful female who was worth the effort. Well, probably this is the time in the sky when we can begin to train the spiritual people how to be the fishermen of mythology, as the Circlet of Pisces is perfectly lining up on the Ecliptic path of the Sun right below the Great Square, the Window to the World, the Cube shaped body of the second Horse, Pegasus. Poseidon or Neptune, the Sea god also loved and married the young Medusa and when Perseus cut off the Head of Medusa, then the drops of blood that fell into the ocean created the steed Pegasus and other drops of blood on the land of the Saharan dessert turned into hissing serpents that so famously represented the hair on the scary face of Medusa.

Thus, Perseus is definitely one of the earliest heroes of the celestial clock. The Northern Brother, which to us is known in astronomy as Perseus, the Rescuer, but also by the names of Mitra, Mithra, Mithras, Gilgamesh, Hunor, Hunaphu, or the biblical Cain, shows up as a the BABOON on the Zodiac of Dendera. The screeching sound of a fearful Baboon, who jumps between Two Trees are the cosmological concept the ancient myth makers desired to record down for us. Than the killing of the magical Bull, the cutting out of the Cedar of Lebanon and the fast rolling Chariot of the fiery Sun lead us to the cosmological TWINS, who heat up the Earth on the Egyptian, Hindu and Mayan depictions. Their famous or maybe infamous hot love affair would be celebrated on February 14th, the Valentine's Day (another 'VA'), the official holidays of lovers. During the cataclysmic earth changes, the tales of the two founding brothers of different nations and the two or three sisters of the same gradually melted into a cosmic union or of this hot romance of the Twins.

In this ancient cosmological model, we shall discover the World Tree, which is the Tree of Life, whose electro-magnetic roots originate in the Heart of the west of our Galaxy. Then we can follow that invisible rope that the Shamans and Medicine Men would tie around their body – and we find the Two Pillars tied with this same strong electro-magnetic rope that reaches all the way to the Sun.

The distinguished and world-renowned Catholic scholar, Professor Mircea Eliade, recognized early on that the Hindu Sacred Trees and thus, all sacred trees of cosmology are connected to the concept of the Axis Mundi, the Axis of the World:

> However the (East) Indians also have the symbol of a Cosmic Tree the Axis Mundi, and this mythological-symbolic is universal, for we find Cosmic Trees everywhere among ancient civilizations . . . it embodies the sacred significance of the universe in Constant renewal of life . . .
> —Mircea Eliade, *Patterns in Comparative Religion*

As time progresses from December to the end of January the Bull changes quickly into the story of the war Chariot of the erupting Sun, and soon after that the Fire Twins begin heating up the inside of the Earth's geo-dynamo. Ancient astrological drawings depict the Charioteer as a person carrying a Goat on his shoulder, and a few little Goat Kids under his arms. That is because the main star of the star constellation Auriga, the Chariot, is the well-known star of Capella, the Goat star that has three Kids born to her. These are three Sun eruption from the Goat headed Devil. Also, to draw attention to the importance of the star Capella – the Chinese begin their New Year in the last week of January, and the classical Greek mythology anointed the Goat Capella as the nanny for the orphaned main Greek god, Zeus. Not the best looking Mother for the main god of the Greeks, but the ancient astronomers, philosophers and talented poets needed to emphasize the celestial position of the star Capella. Capella and the gods are naturally about the important earth change scenarios. In ancient Greece, the Charioteer was also known as Erechtheus, the son of Hephaestus, who was the equivalent of the Roman god **VULCAN!** Now, apparently, Vulcan, the BLACKSMITH (Iron core!), invented the Chariot to be able to shuffle his crippled body around. He was Lame (lame = 'sánta' in Magyar) and He needed the Sled / Chariot to get around. The Christian tradition of Santa is thus, derived from the Pagan believes of the Siberian shamans of the Magyars, Kelta, German and Viking traditions. Since, we are on the subject of Pagan, Heathen, and Barbarian believes, we would like to state that in our opinion, every culture of our Mother Earth believed in an Almighty Spirit God, Great Spirit or some Nature Force that maintains life on Earth. Not necessarily, any of the warring sections of organized religions possess more faith then the Amazon

The World Tree 141

Indians or the Australian Aboriginal people. Our gods are not better than theirs and theirs are not better than ours!

A written form of the Old Cosmic Tradition does not sanctify one story versus the other, especially not when there spirituality and the astronomy are equally represented in both forms. The Creation Legends are about the last time the Earth has gone through astronomical changes and not about a specific human who was allowed to sit by the Creator to observe His creation from nothing to something.

Well, this is how the cosmic tales of the overheated geo-dynamo were kept in the active imaginations and legends of the forefathers,

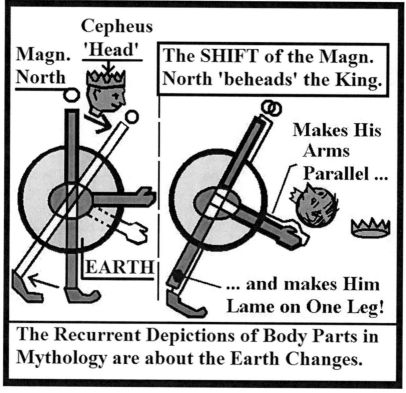

Figure 5-9: Here we show the concept of Beheading the Old King. On the combined Father Earth (Geb) and Mother Earth Model His Arms go parallel and he becomes lame on one leg. After this, he will need a Chariot to be transported around (Drawing by Wm. Gaspar)

who slowly forgot the secret meanings of the volcanically active Globe, but kept the tales recorded and carving on the arts of rocks, animal skins and pergaments.

So, the leg of Mother Earth became injured and the legend of the Egyptian and the Siberian Lame Shaman, Sánta and his Reindeer-drawn Chariot became the basis of countless stories from the east to the west all the way to our times. The name of the Lakota Sioux Native American Indian Medicine men, LAME DEER, just as foretelling - about the specific month of the year when the 'crippling' changes happened then the Greek mythology, the hunter Nimrod, or the Hindu Prajāpati (Orion) whose daughter the DEER called Aurora was born by incest.

The Chariot story is readily present in a number of battle scenes, including the story of Krishna and Arjuna, in the Holy Book of Bhagavad Gita, where Krishna is associated with the Chariot; also there other mythological and religious stories. Most readers of mythology and religions are looking for a historical meaning to the Chariot, but cosmologically the only Chariot that matters is the celestial Chariot of the Sun.

In the Holy Books we encounter evidence of the knowledge of the cosmic events of the Fourth and the Fifth Hours of the Last Twelve where first the erupting Sun causes extreme dryness and then the beginnings of the axis shift is represented by the mentioning of the Polar North position of the Jackal after the clues for the Deer (above Orion) and the Lameness (Axis shift). Without our Cosmic Model these verses would be almost impossible to given logical and astronomical sequences, let alone meanings.

6 *Then the lame shall leap like a deer, and the tongue of the dumb sing. For waters shall burst forth in the wilderness, and stream in the desert.*

7 *The parched ground shall become a pool, and the thirsty land springs of water; in the habitation of jackals, where each lay, there shall be grass with reeds and rushes.*

--Bible, Isaiah 35: 6-7,

Thus, whenever we encounter the mentioning of certain animals

The World Tree 143

Figure 5-10: The statue of the Pioneers in Santa Fe, New Mexico. The statue in the garden of The Cathedral Basilica of St. Francis of Assisi contains a Four Directional Alignment, with the Bull facing to the right, the Sheep facing us, and next to it, is the Goat, lying down. To the left is the Donkey. This is the Astronomy from Christmas to Easter. (Photo by Eva Gaspar)

Figure 5-11: The Sacred Feminine is displayed with this beautiful, Native American woman statue standing in front of St. Francis Cathedral in Santa Fe, New Mexico. She is holding a fan of Eagle Feathers, and the Cross is right below that. Behind her, on the wall of the church, are three pillars holding two arches. On the left upper corner of the church are Two Spirals that to us represent the Two Spiral Pairs of the Galaxy (Photo courtesy of Eva Gaspar.)

The World Tree 145

of astronomy connected with harsh earth changes, we can place them into our Cosmic Model and almost be able to roll a 3-D movie in our heads to exactly observe a series of events happening to Mother Earth. There are even innocent appearing statues around us on town squares displaying the knowledge of a rotating Earth with the properly placed astronomical animals recording down the knowledge of the most important markers of a Cosmic Tragedy. These farms animals radiate a sense of familiarity and security, food supplies in a sometimes harsh environment of the advancing Pioneers of the American Wild West, but the deeper mysteries of these rotating cosmic scenarios hide more horror than any dueling pair of cowboys or flying showers of hissing arrows from the screaming Indians.

One would not expect to find hidden astronomy in statues erected at St. Francis Catholic Cathedral in New Mexico, but the animals certainly represent an astronomical progression of the Bull (January 1), the Lamb (the whole season from Christmas to Easter), the Goat (January 23) and the Donkey (March 21, Spring Equinox).

One would not necessarily expect to find astronomy and cosmology in smaller churches, cathedrals, mosques, or synagogues, but the experience is just the opposite. When one knows the basic Cosmic Model, with its attendant hero and animal mythologies, then every of those holy places is teeming with revealing statues and paintings. Certainly, when someone visits the famous cathedrals of France, Scotland, Germany, England, Italy, the Taj Majal, Angkor Vat, Mayan and Egyptian Pyramids, Chinese Temples, the famous collections of the Vatican, Israel, Greece, and Turkey, a large storehouse of astronomy and cosmology awaits.

Returning to the specifics of the late winter months, we will notice the Chariot of Fire, with the same PENTAGON shape in the astronomical representation, as we saw earlier. Gemini lines up in the middle of February in the sky, if the sky is looked at from south to north in the direction of Polaris. The Beehive would show up in the second half of March. The Polar Bear is Polaris, the North Pole tied to the Pillar concept of the Globe.

Naturally, those playful Bears like to climb trees and eat honey. Now, we know what happens when a golden yellow Beehive is disturbed? Well, the warning is that it might be buzzing with angry

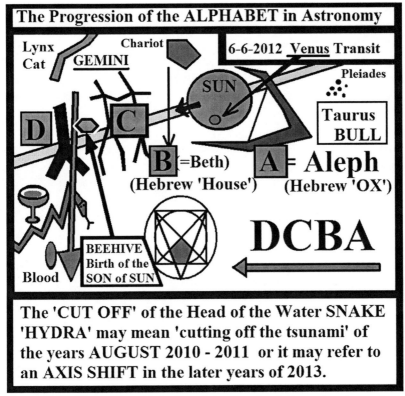

Figure 5-12: The rolling Alphabet of the first few letters also show up on the Ecliptic Path of the Sun beginning with 'A' / Ox / Alpha and ending up with 'D' Delta in the Beheading of Water Snake Hydra. This is where we find the Holy Grail, the astronomical Crater that will collect the Blood of the Sacreficied. (Drawing by Wm. Gaspar)

bees that can assemble and attack from the similarly-colored Sun. From real life examples the Ancients built beautiful Cosmic Tales that are timeless, and even if the precise Astronomy is not understood, the entertaining tales can remain in circulation and cheer up millions of clueless listeners. The innocent appearing children's stories often kept alive the astronomical secrets, such is the case with Winnie the Pooh, the cuddliest little the bear, who lives inside a tree and is always consuming honey from the beehive. Furthermore, he has a few astronomical friends, such as the donkey and a tiger. Is this coincidence or is it keeping the enigmatic astronomical animals alive in the minds of unsuspecting children and adult to circulate the knowledge of the living Cosmos?

Thus, this was one of the many secret ways the Ancients honored the complex cosmological knowledge from about 12,000 years ago. We discovered a lot of, so far unsolved mysteries of the Ancients tales, but we bet that there are much more exact secrets out there.

CHAPTER **6**

The Cosmic Language

*Then God said, "Let there be light" and
there was light.*
— Bible, Genesis 1:3

According to ancient writings, including the Holy Bible, there was once a common language on Earth that everybody spoke. We do not know how true that is, but if a common language existed, it had to be built on the combined knowledge of how the cosmos works. The Ice Age cycle's recurrent natural catastrophes that happened to our Earth over the last 100,000 years of history carved a painful memory on the psyche of the surviving humanity. An obvious need arose for the remnants who survived to propagate the understanding of earth changes. The accumulated astronomical knowledge of these natural cycles was preserved in their art, music, mythology, religion, and ceremonies. The unaltered picture language that we can investigate today is the hieroglyphic picture language of the Egyptians that might have its origins in the pre-diluvian language of the Atlanteans, including the Uyghur tongue. The Uyghur people of the ancient forgotten past, according to Tibetan writings interpreted by Col. James Churchward claimed more than 70,000 year history. During those thousands of years, they reportedly survived repeated and enormous earth changes. Who were the ancient Uyghurs?

The World Tree

First of all, the Uy ('Új') word means 'new' in the Magyar tongue from the Finn-Ugor (Uygur) language family. Thus, the descendants of the original people had to be known as Gor, Gur, Ger, Gár and even Gör. This group, who claims to be the grandfathers of all white people, has two main groups remained in Asia. Neither of them was able to maintain their Caucasian roots in the fast expanding Oriental world. The larger of the two groups are the city living and stationary Uyghurs, and the smaller group is the nomadic Ugors. These people still live north of the Himalayas, where most white nomadic people moved after the expansion of the Hindu civilization. The Uyghurs lost their independence after the 1st World War, but still live in the Uyghur – Xinjiang Province that became an autonomous state of the Communist Chinese Empire. Watching the news nowadays, one can notice that a rebellion is brewing in that part of the world against the ever expanding Chinese rulers.

In the old Chinese history books one can find mention that there were two larger groups of horse riding people, armed with fast and powerful recurve bows over 2,000 years ago that were raiding the western part of the Chinese Empire. The people to the north were tall and light skin, and the people to the south were brown eyed, dark haired and tan skinned. These are the people who they built the Great Wall of China against. In today's ethnic designations, we can think of the northerners as the Viking / Finn/ Kelta and other Germanic races and to the south the Hindu / Hun / Latino and Turkish people.

When history talks about the 'evil Huns', who terrorized the Chinese Empire, those were the descendants of the same Hun warriors who fought with Attila in Europe around 450AD. Their Siberian origin and their successful wars again ancient Greece and Rome are part of history today. Attila, the Hun – the Scourge of God - is considered the founding father of modern day Hungary by the Hungarians, who today call themselves Magyars. The Magyars, who unified the main tribes of the Huns and the Gárs, apparently propagated an ancient language that was based on Cosmology. This mysterious Magyar tribe from Sumer claimed to be the descendants of the Sumerian Magi Astronomy Priesthood. The Magyars promised to lead the Hungarians back into central Europe under one condition. The language, the major offices of the tribal and kingly government, the throne of future Hungarian

Kings, nobles, priesthood and even the foreign attachés all had to be elected only from this one Magyar tribe. During their intense nation making, this tribal union accepted three more tribes from the Alani Persians, also at least one more tribe of Kazárs, who already converted to Judaism.

Thus, after the mainly Latino Huns and the Viking Gárs were intermarrying around the town of Kiev - that they founded in the land of the Kazakhs around the 5th and 6th centuries - they decided to end their unsustainable nomadic ways and settle down. Much before they decided to move to the land of Attila, they attempted to take over one of the large towns in Babylonia, in nowadays Iraq. This happened in the late 600's AD, but that move almost became fatal. Outnumbered and surrounded, they narrowly escaped from the area of the river Tigris under the protection of the night sky. They rushed through via a Pass through the Caucasian Mountains, a famous pass that to this day is called by the Russian people the Bengerski / Vengerski Pass (Vengerski = 'Hungarians' in the Russian language).

After that disaster, they gradually prepared to build up their forces over the next 200 years to reoccupy by force the European land of Attila. This varied group of horseback warriors, who later incorporated some of the Greeks and Romans, Germanic nomads and the farming Slavs of the region, became the legendary Magyars. The neighboring and distantly related countries and of the Germans, Italians and Slavs, who readily received the foreign attachés of the newly formed Christian Hungary, were told that the people who live in Hungary (Magyar Country) were all Magyars and they spoke the language of Magyar.

This Magyar language, just as the Hindu, Turkish, Hebrew, Latin and Arabic apparently possessed important elements of the ancient Egyptian. The pre-diluvian Atlantean and Egyptian languages had to be related to the ancient Mayans and Peruvian from Atlantis.

Thus, major parts of the ancient Uyghur languages, alongside with related parts of Latin and Scandinavian tongues were incorporated into this 'Cosmic' astronomy language. That is why the linguists today cannot place the Magyar language into any group, and that is the reason why we can say words or full sentences in the Magyar tongue that either sounds like Spanish, English, or Finnish, then further on we have words that are identical to Egyptian, Coptic, Hebrew, Mongolian, Arab, Sanskrit, Slav, Armenian, and Slavic. The Magyar Magi perfected the language to the point that almost a computer would have not done better. That is why, the famous 18th century British traveler and politician and ambassador to Asian countries, Sir John Bowring fluent in English, Magyar and four other languages stated something like this; - *'who ever solves the origin of the Magyar language, will be going back to the roots of the Original Language that began with the statement in the Bible that there was first the Word and the Word was God'.*

Well, that is naturally a gross overstatement, since we know that the original Semitic languages of the Egyptians, Hebrews, Ethiopians, Arabs, Turks, Greeks, Latinos and also the Sanskrit, Hindi, Maya, Peruvian, African languages had as much to do with the origins of the languages than the Uyghur or the Magyar. Although, rebuilding of the cosmic roots in Magyar, English and the Latin languages from the fading order of the ancient Egyptian picture language allowed us to solve a good portion of the Astronomy.

Probably, this is why, there are a growing number of Spanish and Basque Secret Societies in Argentina and Latin America today that believe that they should learn Magyar to solve the meanings of ancient carvings in the Americas that are believed to have originated in the time of Atlantis. These sometime valid and other times outrageous claims originated with the eccentric Magyar adventurer named János 'Juan' Móritz, whose story can be found in the book written by the late Scottish entrepreneur and adventurer the great Stan Hall whose book titled *Tayos Gold: The Archives of Atlantis* is gaining recognition. Not many people aware of the fact, that the apparent discoveries of the famed and legendary Gold Library in the caves of the Tayos Indians of Ecuador by János Juan Móritz was partly the basis for the 1972 blockbuster book titled *Gold of the Gods* by Erich von Däniken.

Thus, we write about these things to bring attention to another aspect of history that helped to form our linguistic past. Very few people aware of the historical significance of the 3 – 4,000 year old Scandinavian or Irish Kelta 'Blond Mummies' that were unearthed in the western provinces of China. Even an educational program was shown on TV about 'The Desert Mummies' that was based on the researches and books of number of scholars, including the researches of professor Mair, and Elizabeth Wayland Barber's book titled The Mummies of Ürümchi. Thus, these educational programs detail the findings of 3,000 – 4,000 year old mummies in western China who are well preserved and clearly tall, blond or brownish red headed people. One of the lady mummies named The Beauty of Loulan. She was dressed in buck skin, was buried in the fetal position. Some of these mummies either had an Eagle feather in their hair or a Celtic pointy witch hat next to them. This information likely changes our understanding of the 'native' nomadic peoples' origins from all corners of Asia. This is how Barber writes about the Uyghur people of China.

But the Uyghurs do have a good reason to think of these mummies as among their ancestors, for a noticeable number of the non-Chinese people living in Xinjiang today have blue eyes and light brown or reddish hair a legacy of old intermarriages with the ancient early arrivals from the West,
–Elizabeth Wayland Barber, The Mummies of Ürümchi,

Where Barber maybe wrong, is that there was no need for the 'arrivals from the West', since the white Uyghurs lived in northern India and north of the Himalayas since the beginnings of the current Holocene Epoch. The racist ideas of Hitler, similarly is not at all correct about 'Aryan' people moving to northern India, since that is where they lived originally. The ancient migration routes were implanted first in the desert of the famous Silk Road originating from the Himalayas north and west ward, then later the hoof steps of the merchant caravans crisscrossed the Euro-Asiatic continent in a fashion similar to a spider web. The increasingly farming and ranching stationary life style of the Orientals and some Turkic tribes in that part of the world allowed the Han Chinese to propagate and expand and push the 'pale faces' toward the west. The descendants of the nomadic white Uyghurs and Ugors slowly wondered toward the West to avoid the expanding Yellow Empire of the Chinese. Some mixed with them, became Turkic and Mongols with a stronger sense of their eastern culture having both

The World Tree 153

the outposts of traders and the nomadic settlers and warriors that repainted the faces of the Siberian steppes. Those Uyghurs who mixed with the Han Chinese gave their origin to the Turk people of Asia and founded the great Mongol Empire of Dzhingis Khan.

Thus, the people of the Great Chinese Empire began to be built out of the mixing of the Eskimos from the North with the Nordic Uyghurs, southern Huns and Latino caballeros and the nomadic Turkic tribes. The mixing of these three races (Hindu 'Arians', Hindu Hindis / Latinos and the Eskimos) gave the foundations of most cultures in Asia and even part of Europe and Asia Minor today. That is why we have the Mongolian name of the 'Hero' written 'Baator', very similar to the Magyar 'Bátor' that means 'Brave' and the English 'Bat' that our astronomical Hero, Hercules carries in his hands.

This is how interconnected the ancient languages are, and that makes it difficult to unravel the ancient astronomy from the ever changing creation legends of the newly emerging empires.

We assume that the passing down of the languages happened such, that the most ancient language that came out of Atlantis and Egypt came to us on the Big Boat, the Barque / Barca, Bárka or the Ark after the Flood. The Hindu, Sumerian, Babylonian and other related Flood legends state that there were a lot of people on the original Ark, including those who helped the Hero built it. Yes, we understand that the sacred teachings of the Hebrews only mention Eight people, Noah and seven more, but this we think is the secret cosmological reference for the occupants of the oldest Egyptian 'Boat' the Solar Barque where we commonly see eight deities.

Thus, one of the main Heroes who were highlighted in that 'solar bark' is the Hawk headed Egyptian HERU, (would be spelled 'Jeru' in the Semitic Latin), whose standing body in the 'solar barque' is radiating the rays of the Sun. The Hawk, Heru (Horus) is encircled with the resonating body of a huge Serpent in the middle of the boat. This is Cosmology, just as all the Creation Legends of the nations around Egypt, who either originated or copied these tales. The farther away we are in time from that original astronomical tale, the more we are thinking that the historical renderings of the brave nations around Egypt became entangled and merged with the pure Cosmology of the Pyramid builders. This sort of thinking is the only way that we can open up the star knowledge to everybody without getting entangled in the false tales of ethnic rivalries.

Thus, we think that before the Flood, there was a larger core group that survived with the knowledge of Atlantis and Egypt. Their 'Boat' landed in the Himalayas that are named in the Hindu legends 'Mount Meru' (M-R-), which is the plural name of the Compass in Egypt. The same Egyptian cosmologically oriented 'M-R-'consonants gave us the Mount MoRi-ah ('-ah / ha' means 'the' in Hebrew) designation from the Bible. The Egyptian, Hindu, Sumerian, Babylonian, Hebrew, and Nordic tales in their cosmological essences remained very similar, but the changing landscape yielded different 'historical' sacred mountains as the burgeoning population moved west and east. At the end of the westward directed wonderings of the survivals, the tales of the Hero's universal journey, the 'Last Twelve years resulting in death of the Hero', unintentionally became a gentle floating through the 'Lake Avalon' of the King Arthur. This fine lake was meant to mean the 'watery death' of every Hero who survived the Big Flood. If nothing else, the beautiful and romantic British fairy tale left us with poetic literature, but the brutality of the Flood was not historically represented by Lake Avalon. For our linguistic purposes, we were happy to find another 'VA' in the lake a-VA-lon.

The people and Native tribes, who slowly departed toward the east direction from the Himalayas, for example the Navaho / Dine'é Indians, kept a similar creation legend with the Flood, but this time – when asked – they pointed to the mountains of the Four Corners region of the Southwest of North America as their original mountains of creations. It is perfectly understandable, but for us to decipher the astronomy from the ancient and original tales, the sacred mountains remained the ones recorded down as the secret language of Cosmology.

The concept of the Sacred Mountain / Compass idea remained the main tools of the Freemasons that hide the secrets of the Two Axis of Mother Earth that can close and open at times. The Hundred League nautical measurement difference that we find between the Magnetic North and the Polar North is the Hundred League mentioned in the Hindu myths, the Epic of Gilgamesh and even the Bible. The decipherment and understanding for the basis of that cosmological simile is very important. Without that, we can get lost in the varying creation legends of all

The World Tree

these great nations whose mixed quasi historical cosmological messages through the knowledge of the Compass (Egyptian 'MéR') should stay identical. Only through the uniting perfect knowledge of cosmology can we find our common ancestry. Ethnic separation creates wars, hatred and the ugly efforts of one group trying to annihilate the others to keep 'their version' of the same cosmic tale alive.

The Egyptian word to Measure is also 'MéR' and the Sumerian is 'Mé' or 'Mér'. Then the Magyar 'measure' is 'Mér' and the derivatives of that in Spanish is 'MEdiR' and in English is 'MEasuRe'. Every culture had a reason why they inserted the additional letters and it likely served the cosmic understanding in that particular language. Let us show you that even Jesus Christ, in the 13th century world of the Knight Templars, proud protectors of the road to Jerusalem (serving the Catholic Pope) depicted their Lord with the round Earth and Jesus holding the Compass. Thus, not one great nation or religion should feel inferior to the others, as we all tried to keep the secret astronomical

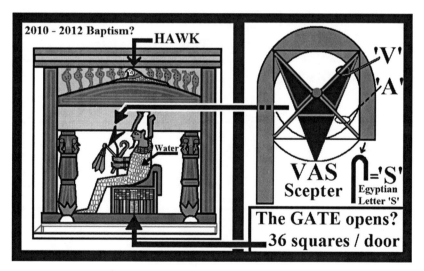

Figure 6-1: The Grand Architect of the Universe, the Freemasonic designation for the Lord depicts Jesus Christ holding a Compass in his hands secretly representing the Two Pillars of the Earth. (Photo courtesy of Acharya S., The Greatest Story Ever Sold, Adventures Unlimited Press, Kempton, Illinois

knowledge alive in our religious lives.

Thus, instead of ethnic rivalry and destructive wars, if we could find the astronomical foundations of the Judeo-Christian, Arab, Hindu, Nordic and Native tribal messages of Cosmology, then we could recognize what the Ancients truly wanted to portray in their tales and what is the unifying message. Once we can establish the basis of that timeless 'sky language', then we will be on our way to also figure out what the bards tried to tell us about how the Cosmos works and what awaits us every 12,000 years. With this basic intention, we further investigate the great linguistic enigmas of those above mentioned languages with the Egyptian as a base, to find out more about the interconnected secrets from all around the world.

Thus, the teaching from this was again the Magnetic Axis shift of the Earth's Geo-dynamo. The AT and the TA syllables survived the test of time. The name AT-tila, AT-lanta, AT-lantis and just as importantly

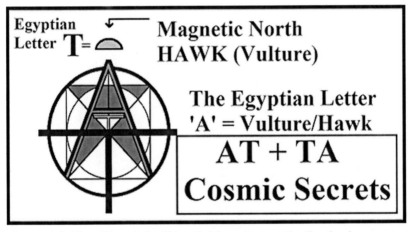

Figure 6-2: The AT and the TA syllables relate to the Earth above water (AT) and below water after the axis shift (TA). Thus, the secret ancient names of AT-lan-TA & AT-lant-IS is partly derived from these two reversible letter combinations. The Egyptian letter 'A' equals the Hawk-Vulture sitting on top of the Magnetic Axis of the Earth - that is Hawk when stable and Vulture when destructive. Then the actual axis or mound shaped half circle picture sign letter represented the Egyptian letter 'T'. (Drawing by Wm. Gaspar)

The World Tree 157

AT-las survived to this day.

It is an exciting endevaour to read and try to decipher words from the telling Egyptian picture signs presented in excellent works, such as the Illustrated Hieroglyphic Handbook written by Ruth Schumann-Antelme & Stéphane Rossini, and also the one by Richard H. Wilkinson titled Reading Egyptian Art. Both books and a few others on the subject are very highly recommended.

We will see even in the next graph, that the word again begins with 'SA' that is the huge Super Sun housing the westerly Black Hole power. The word that ends with 'RÁ' denotes the Egyptian Sun King representing the east. In between these two powerful 'suns' sits our beautiful Mother Earth. Thus, names of people and towns, such as Sa-Ra-i, Sa-Rah, Sa-K-Ra, Sa-Ha-Ra, Sa-qqa-ra, Sa-ma-Ra and Sa-n-d-ra hide complex cosmological enigmas. We think that the knowledge of the Cosmic Model along with the fluency in the colorful Egyptian hieroglyphic picture guide can help us reevaluate our understanding about a common proto-language that may have existed on the continent of Atlantis before the Flood.

The papyrus and the lotus are connected to the Magnetic North and Polar North in Egyptian mythology. The Egyptian 'HA' hieroglyphic symbol shows Three Flower in a Vase with water and loosely that can

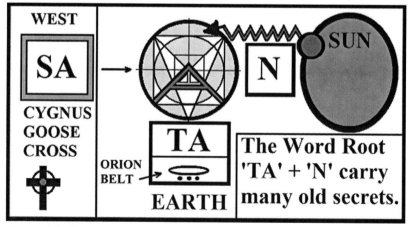

Figure 6-3: The TA-N and the TANOK or TANAKH words are important clues to the Cosmic Sex enigmas that are hidden in religious teachings (Drawing by Wm. Gaspar)

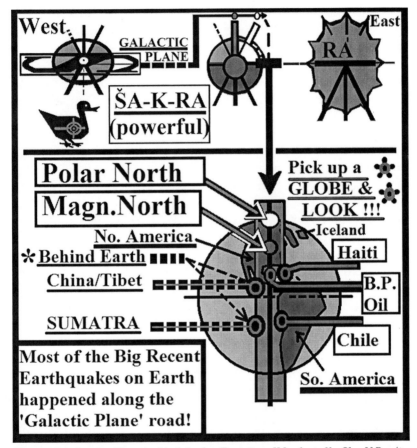

Figure 6-4: In the Cosmic Model, the Egyptian Word spells Sha-K-Ra (in Hindu, Šakra='powerful'). This would be the knowledge of the power of the West (SA) and further of the Galactic Trio. This could be the base word for sacred: sagrado in Spanish, sacrament, also Shah that means the Prince in Persian, then (ba) Sakh means in orderly procession in Hebrew and the Sar is the base for a cabinet Minister also in the ancient language of Hebrew (Drawing by Wm. Gaspar)

be delegated to concepts such as Three, Harem (as multiple wives), Harm, the Arab Haram, the Hawk, Harp, and countless others. These words do not seem to be related, but in the cosmic concept of what happened to the Earth, an argument can be made for a common root

The World Tree

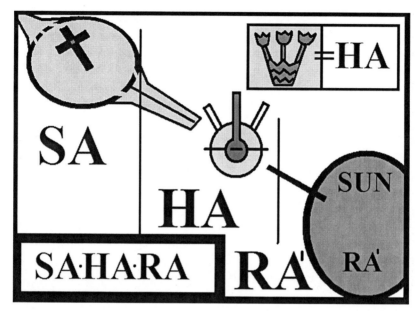

Figure 6-5: The cosmic meaning of the word SA-HA-RA. Even word such as SA-MA-RA that is the "Good SA-MA-Ritan" is derived from this model. The S(z)-A-MÁ-R means donkey in Magyar and it is an important star couple in the Cancer star constellation, and the biblical story of the enigmatic Samaritan word likely relates to the astronomical Donkey (Drawing by Wm. Gaspar)

for the different words and their syllables.

In Hebrew, there are a large collection of words that begin with HA and carry a negative connotation or a related meaning to the space between the two north axes. For example, HA-R-ey Ga'ash means *volcano*. HA-roos means *ruined/destroyed*, HA-tkafah is *attack*, and HAz'har/ah is *warning*.

In the Maya language, the arrow is named HA-lal, and HA-lál in Magyar means death. Are these words just coincidentally similar to each other or do they have common roots?

We have looked at the Egyptian hieroglyphic symbol HA and the next section will bring in the PA, RA, and the KA bilateral hieroglyphic symbols to compare.}

The Pharaoh KaPaRá who is unfortunately written as Khepera or Khepri in English should be known as KaPaRá, since it is made up of the Egyptian biliterals Ka, Pa, and Rá. Once people look at these Egyptian syllables then the cosmic meanings will become clearer. The 'free style' spelling and pronunciation of the English languages is detrimental to the understanding and the dispersal of the ancient Egyptian linguistic wisdom.

The Ka bilateral syllable is the easier one to explain in the earth changes scenario. The Earth axes are depicted as arms of the globe's body, and when those arms go parallel, it is then called the KA sign. Even the word 'KARdio' would make more sense with this figure, relating to the Heart between the Two Arms. Now, since the 'KA' sign in Egypt is made up of the two parallel arms, it is not a far stretch to look for arm words in other languages. Thus, Ka also means the number Two (two arms) in BOTH the Egyptian and Mayan languages!

In Hebrew, for example, the word to be armed (with a weapon?) is KhA-moosh. The Mayan word for arm is KA-B, and the Magyar (Hun-Garian), word is KA-R. Thus, in Hebrew, Mayan, and Magyar languages, the KA initial syllable of the word for arm is present, just as we see it in Egyptian. That is why we need to maintain the knowledge of the letter 'K' and not change it to letter 'C', please! Some of the possible connotations from English to the Egyptian cosmic signs could include obvious words, such as Carry (KA-rry), CA-rgo (KA-rgo), CA-rdinal points (KA-rdinal).

The same thing happened to the 'F' to 'Ph' change. From the original 'Fila-Del-fia' it is now known to us as Philadelphia. If it was so important to keep the town names repeated in the USA from Egypt (Memphis, Alexandria) and Greece (Philadelphia, Atlanta) then we should keep the original spelling as it was intended. One might ask why we are worried about this when we should be thinking about 2012. Well, since language is one of the seven arts and an important way to maintain coded secrets, we feel that it is a serious mistake of new and emerging empires to dilute these ancient root words that were packed full of enigmas.

Obviously, the scholars who started studying those initial pictures and established the basis for the spelling and sounding out of those signs were the French, the English, and the Italians, those same researchers

The World Tree

Figure 6-6: The Egyptian 'HA' symbol (Drawing by Wm. Gaspar)

Figure 6-7: Showing the KA symbol of Egypt, KA also means the number 'Two' in both the Mayan and the Egyptian languages. (Drawing by Wm. Gaspar)

whose languages are the most affected by the language altering processes of the past. Thus, in the English language, the letter O can be pronounced FOUR different ways, depending on what other letter is near the O, or sometimes just a random pronunciation without a worthy rule. Another example is the letter 'I' represented by the 'feather' as the weight of the cosmic wind. Certainly, the English only pronounces 'I' as 'áy' sometimes, thus resulting in the misspelling of a number of Egyptian words. I do not have to emphasize that almost everybody else around the world pronounces the 'I' as letter 'I', as it would be pronounced properly even in the English words, in instances, such as 'in', 'thin', 'limp', pimp and others. We understand that a process to internationalize the Egyptian

The World Tree

hieroglyphic picture language is under way.

The Spanish word Acaparar ('Akaparar') means *raking it in* or *hoarding* in as in grabbing things with the Arms at the dinner table. The tool Kapa means a *Hoe* and the verb Kapar mean *scratching* in Magyar. Thus, the Egyptian KA symbol that is showing the parallel Arms has a meaning in Maya, Hebrew, English, Spanish and Magyar conceptually. Since, the English language mispronounces and misspells these essential cosmic words and the Egyptologists main international working language is English, therefore we commonly lose the actual meaningful cosmology hidden in these symbolic words.

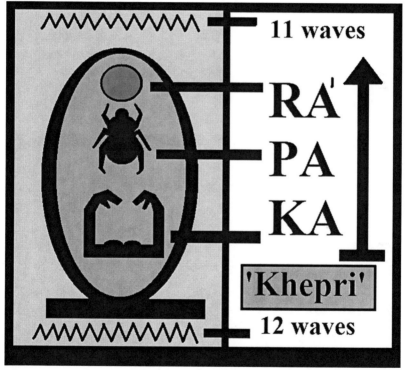

Figure 6-8: The Pharaoh's (FaRaO / FaRaOn) name written as Khepri or Khepera in the English version, which is, naturally, the wrong spelling! The English language commonly changes the 'A' sound to an 'E' sound, but please do not change the spelling! The hieroglyphic symbols should be read as KaPaRá. (Graph by Wm. Gaspar)

We hope to provide a basic working model for the study of the ancient secrets, sacred geometry, and astronomy where the animals, tools, and godly heroes would make perfect sense in a scientific understanding. All the wild and domesticated animals that became associated with the devil, such as the serpent, ram, goat, and bull make perfect sense in our astronomical model.

In this last figure, we can observe that the Earth's geo-dynamo is spelling out KO, or more properly KOR, although the English translation of the Koptic Egyptian pronounces it as Ku. Certainly, we have ample reasons to believe that the original and proper name of the **Arm holding a Whip** is 'KOR'. The words that could relate to that ruling concept of the Earth's KORe is the KOR-a-zon, Heart in Spanish, The 'KORonary' arteries in the heart also relate to the KOR and connects the Heart (KORaZoN). Another famous ruling item would be the King's KORona ('crown'). This Iron KORe of the Earth is closely tied to the KO-RA-ZON of the Earth and probably also tied to the Galactic Center's KORe. Thus, when the Iron KORe shifts, things

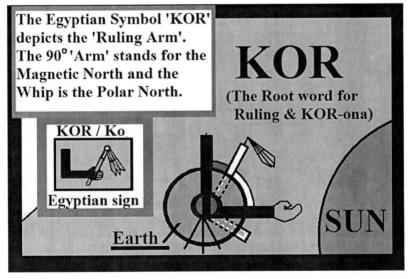

Figure 6-9: The KOR Egyptian hieroglyphic sign relates to the Rulership of the Pharaoh. The official meaning or spelling is not certain, but used by the Egyptologists as Koo / Ku / Ko / or Kor (Drawing by Wm. Gaspar)

turn bad, and what happens to iron when it decays? It becomes KOR-RO-sive.

Returning back to the KOR sign, we can notice that there is a Whip in the hand of the Pharaoh. In the ancient Magyar language the KOR word root yields several similarly spelled and conceptually related words. Thus, the Magyar word *KOR-bács* means *whip* and the Koptic equivalent, spelled KUR-bash, also means whip. The Government in Magyar is written as KOR-mány and to 'keep them in check' means *'keeping them in KOR-da'*. The Magyar word for *disease* is KÓR and the Hospital is KÓR-ház. Another important Magyar word with KOR is 'Age' (Kor) as *someone's age* and *a long term age* ('chronic') and KOR-nyikál, that is 'singing bad' (the news of the ages –'Chronicles' or the Greek 'Chronos'). That means to us that when the Rulership of the Pharaoh has changed, it happened at the end of an age and the news is that it was violent and then disease processes followed.

The Egyptian pharaoh is depicted with the Royal Scepter or Staff, which by now we understand to stand for one of the Earth's Axis. Let us see our next figure:

The Egyptian pharaoh is shown with the "Was Scepter" or Iron Rod that is the symbol of the Magnetic North. The Whip is the

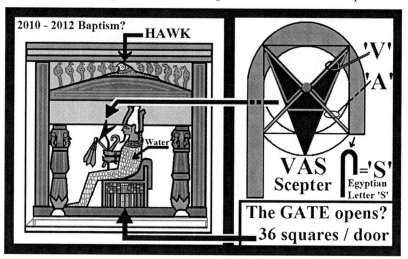

Figure 6-10: The pharaoh with the 'VA-S' ('was') scepter and a KOR-bash whip (Drawing by Wm. Gaspar)

representative of the Polar North. When the ruler holds his or her arms across the chest, that means the Magnetic North Axis shifted parallel to the North Axis. It is not easy to find the linguistic enigmas and the clear cosmology in the meanings of the Egyptian picture symbols, when modern languages are changing around us, although we believe that a disciplined international approach could yield us valuable results.

The last secret linguistic examples we would like to present is a painting by Hans Memling that contains numerous Cosmic Enigmas.

Hans Memling famous painting in Figure 6-11 does not yield the colorful images it deserves here in a Black and White media. Let us

Figure 6-11: The Seven Joys of Mary, a famous painting by Hans Memling is the 2nd secret 'TA' we are presenting from the Christian arts and it is not the last one. Kept in the Alte Pinakothek, Bayerische Staatsgemaeldesammlungen, Munich, Germany (Photo Credit: Bildarchiv Preussischer Kulturbesitz / Art Resources, NY)

The World Tree 167

talk about first the beautiful colors displayed in this painting. The Person who stands in the left side of the gate above the kneeling King or Magus wears a bright Red Robe. Below him the kneeling Magus himself is dressed in a Black Robe. To the left of him, the African King of the Three Magi has a White Tunic over his shoulder. So far, we are not surprised, since the Black King represents the heat of South direction, phenotypicaly at least, thus the White color of the Polar North that is our mythological South. The Baby Jesus is shown as a hue of Yellow color to secretly tell us about the East direction from where the Son of the Sun of every culture would be born from. The Virgin Mary wears the Blue color of the Sky and the Green grass in front of them represents the bountiful Earth.

The main participants of this Nativity Scene are positioned in a gate that seems to end in a CAVE that would be reminiscent of the Dark Rift 'cave'. It is in a mythological mountain that the Hindus called Mount Meru and the Bible renamed Mount Moriah. The Rock outcroppings on that mountain, reveals to us the ROCK that comprises the Core of Mother Earth, the Rock on which Christ later was Judged and the Cave where He was laid to rest before Resurrection. On top of the Gate we again can clearly observe the roof displaying the secret 'TA' of Cosmology standing for the SACRED FEMININE. So, now that we have found the Sacred Feminine in this Nativity painting, we are curious to detect the SACRED MASCULINE! Most religions that believe in a Male God would find the concept of this Sacred Masculine redundant, but in our Cosmic Model of the Unifying Male and Female forces, it is not as obvious. Since, we have already introduced the name of the original Male Fertility God of Egypt, MIN it is not so surprising to find MINisters of the Faith painted below the MIN-aret shaped building standing for the phallic symbol of Cosmic Union. The design of the painting from 1480 AD clearly demonstrates a very precise knowledge of Cosmology even during the Dark Ages. On the front lawn to south of Baby Jesus we find the Dog, who astronomically represents the Dog Star Sirius.

The knowledge was there hundreds of years ago, it is still there today in the possessions of a handful of secret keepers. The parts of the puzzles are there, we provided the foundations for a more meaningful Art History class, and now it is in the hands of the seekers to benefit from the knowledge of Cosmology.

CHAPTER 7

Ancient Mayan, Hindu, and Egyptian Roots

What will be comes to be, and no one can prevent it; what has been, has been.
—Hindu Myths

Is there any validity to reincarnation, and to the antediluvian existence of Atlantis? Nowadays, an increasing number of scholars and average people are coming to believe in these ideas. The number of books written on these subjects has grown at an unprecedented rate in the last few decades. Some of these books claim the knowledge has come from historical writings from ancient Greece and even from Tibet and other Asian monasteries. Thus, we have information about the ancient civilizations that once existed. To maintain an unadulterated message without the effects of the shifting cultures and changing languages and religions, the ancients, including the Egyptians, developed a system and carved their message onto stones.

As we mentioned earlier in Plato's legend of Atlantis in his book, Timaeus, along with Herodotus, and other Greek philosophers' works, the interpretations of harmonic numbers found in the Hindu temples and the Egyptian and Mayan pyramids, psychics and occultists work, and the secret teaching of the Kabbalah, Gnostics all point to great civilizations existing prior to Noah's Flood. There may even be evidence of repeated floods, perhaps several times over the last 100 - 200,000

The World Tree 169

years. If we only read our representative holy books, the messages would be less clear, but with the interpretation of the gestalt of all sacred writings since the beginning of civilizations, the communication crystallizes out.

Drawing from the work of Mr. Schwaller de Lubicz, one of the foremost Egyptologists of our time, John Anthony West, demonstrated in his book Serpent in the Sky, that the Egyptians of millennia ago were a much more evolved society of thinkers than mainstream scientists had allowed. There is no evidence of a slow, gradual evolution of civilization in Egypt. Rather, it is a culture that displays a high level of complexity in the art of math and related sciences, from the beginning of its existence. With the help of Boston University's Dr. Robert M. Schoch, West was able to show that the water damage to the Great Sphinx of Giza was at least 9,000 years old, or even much older. This is now the accepted age of the Sphinx, or LION, reestablished scientifically, and correcting the previous, erroneous assumption of about half that age.

It thus seems very possible that the Great Sphinx was constructed over 12,000 years ago to commemorate the beginning of the Age of Leo, prior to the devastating Flood. It had to be done before the Flood when a technologically capable and advanced civilization was still in existence. If extensive water damage is observed on the Sphinx, then it had to be constructed prior to the Flood and the subsequent raining. After the Flood, the struggling human survivors were so few and limited that building a huge and useless structure was the last thing on their minds. Thus, we reason that the Egyptian culture was the most ancient and may be even was more advanced technologically than ours. When a Deluge comes every, let's say 12, 000 years, then the tallest mountains are covered with the Flood water. As most of us know, water is possibly the most destructive force when it is on the move. If a large tsunami followed a world burning, then a great Deluge would happen as we read the ancient manuscript, there would not be any evidence of the great technology of an advanced civilization a few thousand years from now. First, everything burns then the water washes it away. Slash and burn! It works for humans and it works for the Creator.

Mr. Graham Hancock, in his book, *Fingerprints of the Gods*, argues the same point about the antiquity of this great monument in Egypt;

on page 345, he writes:

The first and lasting impression of the Sphinx, and of its enclosure, is that it is very, very old – not a mere handful of thousands of years, like the fourth dynasty of Egyptian pharaohs, but vastly, remotely fabulously old. This is how the Ancient Egyptians in all periods of their history regarded the monument which they believed guarded the "splendid place of the beginning of all time" and which they revered as the focus of "a great magical power extending over the whole region."

Furthermore, he claims that the "Sahara desert is a young desert, and since the GIZA area in particular was wet and relatively fertile 11,700 to 18,000 years ago," we should consider placing the time of construction between those dates. We agree with these scholars that it is more believable that the fourth dynasty's pharaoh, KHAFRE (CHEPREN) took on the symbol of the Sphinx to show his people some sort of interconnectedness with a powerful message of the monument that may have already existed before his time, than assuming the Sphinx was simply constructed in KHAFRE'S era. Most of these pharaoh gods may only be astronomical tales. Maurice Cotterell writes in his book, The *Tutankhamun* Prophecies, the following:

With the shafts of the Great Pyramid they pointed the way to everlasting life, and rebirth in the stars . . . the worship of Apis the Bull and Amun the Ram concealed the secrets of the heavens, as they buried their dead beneath the Milky Way.

This quote directs our attention to the two main animal symbols worshiped by the Egyptians. The Bull and the Ram make perfect sense in their wise insight into the secret of the heavens.

The ancient Egyptians developed a very sophisticated system that applied a model that in one page stated the Trinity power relations of the sky. In this model, they stated that the western sky controlled the solar system, and the Sun held steady the two axes of the Earth. They managed to manufacture, out of this one graph inspired by

The World Tree

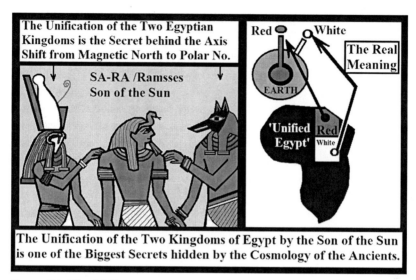

Figure 7-1: The FORM to make Phoenician letters and Egyptian hieroglyphic symbols represented the Knowledge of the Heavens and Earth.

sacred geometry, a number of letters of their Phoenician alphabet, their hieroglyphic symbols, and a number of advanced ideas. Our illuminated astronomy priests used that ONE PAGE system to develop the Heads of their animal gods, that each meant an important function in cosmology. So, how did they do it?

What we are showing here is an absolute first. This to us was a revolutionary discovery, that this one Form was able to contain so many vital pieces of information, all derived from a model that is carved on the tombstone of some US presidents who were the founding fathers. Did they know what it actually meant and what tremendous information it carried, or did they just used the sign to demonstrate the fact that they belonged to the secret organization? Why did this same upside-down, five-pointed star symbol, when encircled, have an association with devil worshipping? The connection between the two is fuzzy for us. Likely the pentagram was demonized so the average folks would stay away from it and not decipher what it means.

One would wonder if they had this very important FORM (=MIN-TA in Magyar), such as a tool and die making device that could press out hundreds of different forms of very worthy information about the state and interrelatedness of the cosmos, then why would they hide it from us.

Well, they did just the opposite! They kept the FORM, **the main secret** in a simplified way, in front of us in PLAIN SIGHT. Some parts of it they placed on the walls of our Lodges and some they printed onto our Money so it would allow us to hand the sacred information to one another without knowing. They placed the different parts of the 'forms' on flags, medals, clubs and company logos and they even named the leading cabinets of our political & religious representatives by those secret names. Now, how many people even in the so called 'secret societies' know these? Not too many!

Certainly, this was not exclusive to the American monetary system;

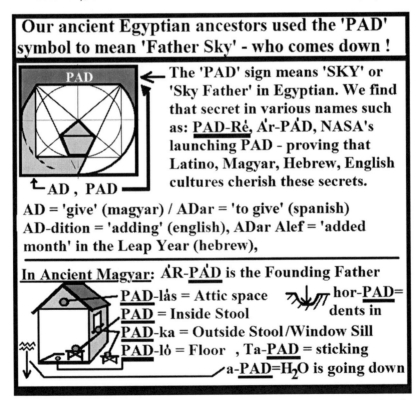

Figure 7-2: *Several different Egyptian hieroglyphic symbols are clearly shown to be derived from this Secret Formula (Drawing by Wm. Gaspar)*

The World Tree 173

we could see parts of this Mono-myth on the pesos of the Mexicans and other countries around the world. One cannot even take medicine, buy a TV, or drink a shot of brandy without running into the blatant advertisement of these cosmological wisdoms as emblems of company logos. This subject could fill another book volume. The keepers of the secrets were so sure that nobody could crack the knowledge of Cosmology that they kept it, and still keep it openly in front of our eyes. We can ignorantly stare at those logos and symbols with a blank face without knowing what they should mean to us cosmologically. Brilliant!

All they had to do is to present the secrets to us in a sexual way and demonize some of the signs, such as the upside down five pointed star, the pentagram so no pious seeker would connect the dots. Certainly, it is okay to make horrific movies about 2012 —nobody would believe it anyway, until they see the wall of tsunami approaching us in actuality.

We would like to introduce the Egyptian animals that will appear on our secret. All these simple-appearing animalistic caricatures hide a superb knowledge of the geo-dynamo in the solar system and in the Milky Way Galaxy!

Therefore, we start understanding that there are **THREE DYNAMOS** in our Milky Way Galaxy. The most powerful in the west

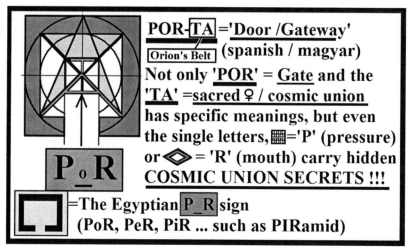

Figure 7-3: Examples of the Egyptian Hieroglyphic symbols and their meanings (Drawing by Wm. Gaspar)

is called 'SA' mythologically. The two pillars of Mother Earth formed the letter 'K' and the Sun was the Egyptian Solar God called 'RÁ'. This 'SA-K-RA' base delineated the linguistic bases of the 'Trinity Forces' of the Galaxy. When the Magnetic North axis of the globe moved, the middle 'K' became a different syllable. Several major techniques evolved to show these changes, most of which we will not detail in

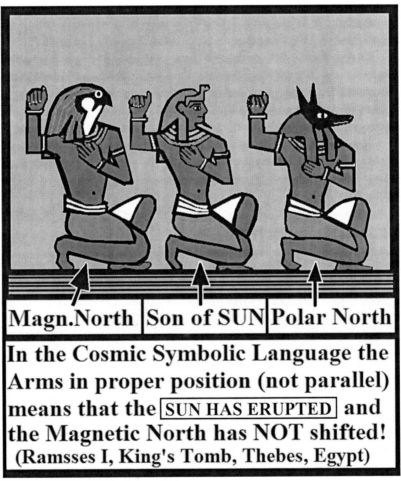

Figure 7-4: Horus, Ramses I, and Anubis kneeling in an Honor or Swear position tells whether we reached the center or not and whether the axis shifted or not (Drawing by Wm. Gaspar)

The World Tree   175

this book. The different syllables that described the movement of the axis of the Earth became 'VA', 'TA', 'MA', 'MO' and others. Similarly, the stable 'Rá' syllable denoting the Egyptian Sun became the letter 'N' or 'NA' to indicate the changes in the erupting Sun. Every major concept earned its own syllables in this complex Egyptian picture system. Naturally, the 'Del', 'De', 'Di' syllables became connected with ideas that connected with delicate 'De'-generate', 'Des'-tructive and 'Dis'-ruptive ideas in this cosmic language scenario.

The Dy-na-mo in the Heart of the Galactic Super Sun Bulge is a

Figure 7-5: The Maya warrior on the walls of Uxmal displays the same angles for the ARMS as its Egyptian counterpart. We have not seen anybody detail a very close correlation of the Maya and Egyptian symbols. We have not seen the correlations of these arm positions to Cosmology, either. From the Mayan city of Uxmal, Mexico (Photo courtesy of William Gaspar.)

Spiral Pair that, in this case, is marked with the Cross. At other times, the main westerly dynamo was identified with the Goose or the Face of the Bearded Pharaoh. In the case of the Mayans in Uxmal, it is also shown with the more scientifically proper Two Spirals. The secret meaning of the Braided Hair was the tool of the nomadic people. They did not build elaborate castles, thus they had to design the astronomical reminders as part of their hair style, clothing and religious regalia that they could carry with them on their constant journeys. Now we begin to understand why the Keltics, Native Americans, Vikings, and Chinese wore the long braided hair parted in the middle. The arm positions of the heroes, priests and rabbis became another method of maintaining a working knowledge of cosmology.

All these animals have specific meanings, as we know, but the **POSITION of their ARMS is always a clue to where the AXIS is!** When the right arm is in a 90 degree angle, it means the Magnetic North is still standing in its place. The left arm is usually in an approximately 24 degree angle, and it is placed on the Heart! That, to us, means that the Two Axes of Mother Earth are still in place. The ancients understood that everything is energy, created out of vibrations and differing frequencies. This ancient knowledge is being rediscovered by today's science.

One observation that we arrived to in this system is that even the English language displays the knowledge of this Cosmic Model with words that mean different things, but in the model they are connected conceptually. These English words should not be spelled very similar as they are only one letter different than their cosmically related concept. Examples of such would be; H – EART – H (Heart / Earth / Hearth), H-ARM, C-O-RE, F-Light and others. Unfortunately, we do not have the time and space to venture into this finding in this book.

We feel that our revelation about the Arm Positions of these mythological figures should help understand the ambiguous science of interpreting these gestures. The famous line 'walking like an Egyptian' is well known and popular, but until now nobody provided us with any clues what they might mean.

The famous spiritual leader and legendary Mayan King Pakal is not an exception of that rule. On the next picture we can observe Lord Pakal. He is standing on a couch that is decorated with beasts. He is

The World Tree

Figure 7-6: Lord Pakal on a couch displays the cosmological arm positions. In essence, with little minor alterations on the picture, one could see a very similar choreography of a dancing or posing deity from the Hindu mythology. (Picture by Wm. Gaspar)

demonstrating the Arm Positions of the Earth's axis. To go even a step further we could bring in statues from India or Cambodia's Angkor Wat to demonstrate the same cosmological messages from a third continent. We venture to say that at a very ancient and basic level – in all currently existing cultures and civilizations, from the fairest to the brownest – we are the descendants of one of those or a complex

This old Mayan King has both of his ARMS tied in a parallel fashion, referring to the position of the poles. The King is also BEHEADED on this picture that has a very specific message relating to the AXIS SHIFT. The Mayan sacred number of 20 is encoded in the tiles of the Temple wall. The Maya has a lot of Egyptian type of Cosmology. (Uxmal, Yucatan, Mayaland)

Figure 7-7: The Mayan arms, parallel and tied, are close to the Egyptian cosmological symbol. The Sacred calendar count of twenty is also present. The fact that there is no head on this figure is not a mistake. It is a beheaded cosmic figure (Picture by Wm. Gaspar)

combination of those three major cultures from a long time ago.

Different theories abound, but the Mono-Myth places all of us back to the cradle of civilization. How long ago, where, by whom, it is difficult to say, but the proof places all of us at One Glorious Prediluvian Past and into One Root Race!

On the Mexican money called Peso, one can see an Eagle, or maybe Hawk, picking out a Serpent from the Water. Since we know that the Swooping Hawk or Swooping Eagle represents the Magnetic North, then it is not a surprise that the Magnetic North / Hawk swoops toward the south to pick out the Water Snake, Hydra, from the River. This is the same legend, but a different continent. One would initially assume that the Creation legends of the Mayans and the Egyptians do not stem from the same root, but not only the animal symbols are very similar, but the angles of the arms of gods depict the same degrees.

Figure 7-8: The Eagle or Hawk picking out the Water Serpent from the River is the same idea as that of the Arms going Parallel (Drawing by Wm. Gaspar)

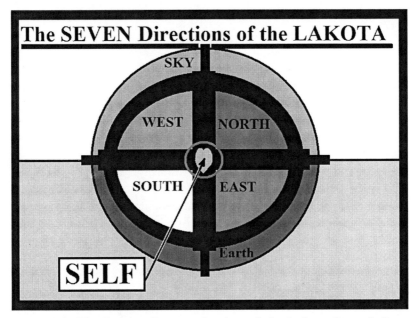

Figure 7-9: The Four or Six sacred colors of the Lakota Sioux Indians (Drawing by Wm. Gaspar)

Even their headdresses and arm bands resemble each other. These two great, ancient civilizations had to have the same root or they were in contact early on.

The Egyptian, Mayan and the related Lakota Sioux Native American Spirituality became our teachers. The intense and secretly cosmologically driven ceremonies the Lakota Sioux Indians made a huge difference in the current understanding of what the Ancients knew about the workings of the Cosmos. First of all, their prayers always begins to the West direction acknowledging the scientific fact that the main energy source that turns our Earth toward the East is definitely derived from the west. The west is where our Sun goes down and it is where our powerful Black Hole is located and they marked the West direction with the Black color.

They carefully placed the foundations of Cosmology into their directional colors, stories, songs and ceremonies. One of their main ceremonies is called the Inipi sweat lodge ceremony, where the tent-like structure they pray in is a half-globe shape. In their teachings that globe shape represents the spherical Mother Earth above the Horizon. Most revealingly, the hot stones that are carried into the lodge from the fire pit located to the East of their lodge represent the hot core of the inside of the globe. Even more amazingly, one can notice that the stones are placed into their positions by Deer antlers. The Deer symbol again connects the Core of the Earth with the upper part of Orion where the Deer is located mythologically. What a simple and beautiful way to preserve an important knowledge about the Geo-dynamo of the Earth. As participants in these ceremonies, including the amazing Sundance, we can honestly say that these ways of the Native Americans are unadulterated vehicles to save the knowledge of Cosmology for thousands of years!

The four day long Sundance ceremony requires the dancers not to eat or drink for four days. That is likely a reminder for the Famine and occasional lack of water, likely not any different than the Catholic Lent or any other religious fasts. The four days would consist of 96 hours, which are the same exact numbers that are found on the tomb of King Tut and hidden by Lord Pakal. In our opinion this number 96 represents the sacred 96 pairs of Venus Transits needed to complete a Half Precession found in the Ice age cycles. The very emotional piercing of the dancers by a rope to the centrally located Sundance Tree and the subsequent breaking away in four trials to us represented the cosmology of the earth changes.

Needless, to say that before the Tree was cut out to be transported to the Sundance circles - even before the dance began - a Virgin girl was needed to find the sacred tree and she was the first one who struck at the trunk with the axe. When the Tree was finally placed into a hole in the center of the Sundance circle a HEART of a BUFFALO was buried beneath the Tree. This equaled in cosmological importance to the Spanish and Mexican toreadors killing the Bull at the shoulder blade that represented the east, the Pleiades, the Heart of the Bull, thus the direction of the Sun! A much longer list of Lakota traditions associated with the Sundance could be brought up as the wisdom to preserve the knowledge of Cosmology and Earth changes. The most important initial realization was that the North (Red) and the South (White) colors of the Lakota Sioux Indians agreed with the colors of the north and the south of the Egyptians. This was one of the main findings propelling us toward developing our 4 Directional Cosmic model.

We'd like to point out an important fact: the Egyptians, the Jews, and the Phoenicians all lacked vowels in their alphabet. Therefore, the name of King TUT or THOTH was translated out of the original TWT. Since it is not easy to pronounce TWT, the word was made into TUT by the Egyptologists. Why is this an interesting piece of information?

Figure 7-10: *The Unification of the Two Egyptians Kingdoms of the North / Red and South / White. (Graph by Wm. Gaspar)*

Well, as we recognize the T to D change in a number of languages, for example, the German GOTT to the English GOD, similarly, there was a change from Egyptian T to the Hebrew D. At least, this is what the Egyptian-born author Moustafa Gadalla claims in his book, *TUT-ANKH-AMEN: The living image of the Lord*.

Since the ancient languages didn't have short vowels, the first element of this king's name was always written as "TWT" i.e., with three consonants. For some mischievous reasons, the middle consonant letter was changes to the vowel "u" by some Egyptologist. When "TWT" is written, in

The World Tree 183

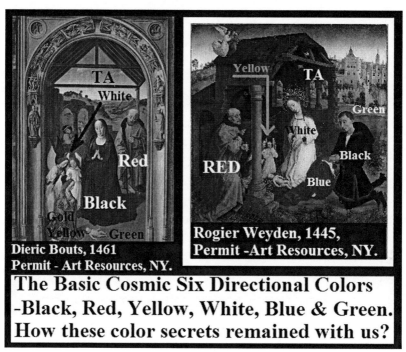

Figure 7-11: A colorful example of how the recent religions employed the same Cosmic Directional Colors as we find with the Lakota Sioux Indians, Mayans and Egyptians (Drawing by Wm. Gaspar)

the equivalent Hebrew alphabetical characters, it becomes "DWD". When "DWD" is pronounced phonetically it becomes "DAWOOD" which is the Hebrew name for "DAVID."

Thereby, Gadalla claims that the stories of King David are Hebrew translations of the accounts of the Egyptian King TUT (TWT). Although, the ethnicities in that region are historically not clear at all, thus the original Egyptians could have been the Semitic tribes themselves. If this is true, then by reading the Torah, Tanakh and the Bible we should receive an important version of the oldest rendition the Egyptian Cosmology. Regardless, this is how Gadalla writes:

We shall prove that basically the biblical account of King David

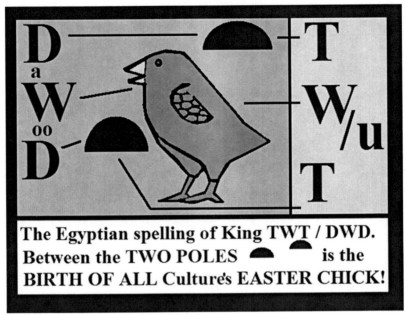

Figure 7-12: King Tut of Egypt (Drawing by Wm. Gaspar)

match precisely the war accounts of TUTHOMOSIS III.

Not having vowels originally in those biblical names present an interesting question. If we just use the J-S from Joshua then where does it lead us to translate JoS-eph, JeS-us, JeSS-e and a number of the other biblical characters? Also, let's not forget the Greek hero Jason.

Please allow us to return to mythology. The internationally known writer, Graham Hancock, author of *Fingerprints of the Gods* quotes from another book on page 243-244. He wrote:

The myths not only describe shared experiences but that they do so in what appears to be a shared symbolic language. The same literary motifs keep cropping up again and again; the same stylistic "props" the same recognizable characters, the same plots.

According to Professor de Santillana this type of uniformity suggests a guiding hand at work. In Hamlet's Mill a seminal and original

The World Tree

thesis on ancient myth written in collaboration with Hertha Von Deschend (Professor of the History of Science at Frankfurt University) he argues that: When something found, say in China, turns up also in Babylonian astrological texts, then it must be assumed to be relevant if it reveals a complex of uncommon images which nobody could claim had risen independently by spontaneous generation. Take the origin of music. Orpheus and his harrowing death may be a poetic creation born in more than one instance in diverse places . . .

Connecting the great universal myths of Cataclysm, it is possible that such coincidences that cannot be coincidences and accident that cannot be accidents, could denote the global influence of an ancient, though as yet unidentified guiding hand? And might not that the same have left is ghostly fingerprints on another body of universal myths? **Those concerning the death and resurrection of gods and great trees around which the earth and heavens turn . . .**
—Graham Hancock, *Fingerprints of the Gods*

Could we translate the mysteries of the Bible with the emerging knowledge of Egyptian cosmology? So, what are the symbols of the Universal Mono-myth that are widely shown, but their meaning is not fully comprehended by the masses? One of the seemingly primitive statues commonly found in Egypt and also in some of our city squares is the obelisk. The obelisk is likely a phallic symbol of ancient origin, but what does it mean?

There are famous obelisks in Egypt. There is one from Luxor Temple that is now erected in Paris. Then, there are the obelisks of Tuthmosis I and Hatshepsut at the Karnak Temple and others. There is the equally well-known Washington Monument, an obelisk, in the capital of the United States. Even in the relatively small town of Santa Fe, New Mexico, there is an obelisk in the middle of the town center. Buena Vista, a tiny village in the state of Colorado, has an obelisk. Furthermore, the Muslims have Minaret's that resemble obelisks. Are these obelisks all over the world representing a phallic symbol of the omnipotent cosmic force of our male Creator? A lot of religions and spiritualities from different parts of the globe do not necessarily subscribe to a male or female creator, but rather believe in an almighty

Figure 7-13: The obelisk in Santa Fe, New Mexico (Photo by Wm. Gaspar)

The World Tree

that is both. Thus, what is the need for a primitive phallic symbol in a civilized world if not the knowledge that, at a predetermined interval, the natural rhythms of our Galaxy require that an eruption of the Sky Father / Sun God will impregnate Mother Earth with an embryo who will rule the New Ages? The designation of the Arab MIN-aret is interesting, because the Egyptian god MIN is the best representative of the phallic worship, therefore we think that the MIN syllable is intended to connect the structure of Muslim prayers with the male god. If most of our pious imams would know that their daily prayers are sang out of the 'corona' of a phallicly designed minaret then they might humble themselves. Although, the blast of a suicide bomber might be a feeble effort to reenact the 'rapture' the male god felt. This is how the Hindu Myths write about that special moment:

As the heat of passion came to the king for his enjoyment, heaven laid aside on the ground the bright seed that had been spilt. Agni caused to be born the blameless benevolent group of youths and made them great . . . Heaven is my father, the engenderer, the navel here. My mother is this wide earth, my close kin. Between these two outstretched bowls is the womb."
—Hindu Myths, *Rig Veda*, Penguin Classics

Thus, in the above quote from the Rig Veda / Hindu Myths, one can find assurance that the legend is about a Sun King and Mother Earth, thus we are truly the products of both a masculine and a feminine entity. The mention of the Fire God, Agni, is part of every sacrifice and re-creation.

When the linga [Hindu phallus] had been established again for the welfare of the worlds, those excellent sages became intent upon the dharma (order) of the classes . . . again and again the universe, moving and still, is burnt by Agni. The supreme purification of this entire universe is to be accomplished by ashes; I place my seed in ashes and sprinkle creatures with it.

Thus, we assume that these similarly based cosmic sex ideas

representing the earth changes originated a very long time ago.

The Zodiacal signs themselves date from extreme antiquity for Professor Alfred Jeremias found evidence that a Sumerian clay tablet in the Berlin Museum (Vat 1847) begins a list of named zodiacal constellations with the name Leo (the Lion), which projects us back to 11,000 BC or thereabouts –the time when the equinox point was in the constellation Leo. This suggests that the zodiac had been devised even before.
—Allan and Delair, *Cataclysm!*

Our conscious human history is much longer then we learnt in our history classes. When further studies will emerge, we will not be surprised if the numbers will be counted in the millions of years.

CHAPTER **8**

The Crocodile Dragon of Mythology

Then I saw another beast coming up out of the earth, and he had two horns like a lamb and spoke like a dragon.
—Bible, Revelation 13:11

One of the most well-known and revered ceremonies of the Egyptians was the opening up of the mouth of the Crocodile. First of all, who closed the Mouth of the CROCODILE in the first place and why is it so important to open it? When one stares at our solar system from the westerly center with the Sun behind the Earth and imagine that the round globe is the face of a celestial clock then the Mouth of the Crocodile is the Two Pillars of the Earth between 12 o'clock and 1 o'clock. On the same face of the clock the Mouth of the Whale or the Great Fish would be positioned in between the Solar Plane of 3 o'clock and the Equator of the globe at 4 o'clock. The closing of the mouths of these beasts naturally happen simultaneously with the right upper hand quadrant of the face of the clock from noon to three moves an hour ahead.

As we intently searched for a model and an exact meaning, we began to look for the Crocodile and Dragon mythologies of past

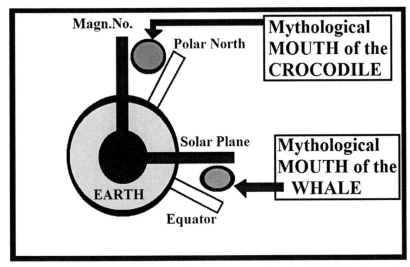

Figure 8-1: The Mouth of the Crocodile and the Mouth of the Whale on the imaginary face of the Celestial Clock (Drawing by Wm. Gaspar)

cultures. Not only we hoped to find an astronomical sign named Dragon, but we also hoped to find it between the Magnetic North and the Polar North poles of our globe. We did encounter the Dragon in the sky between the Two Pillars and we also found an exact period of the year by applying the very useful Star Talker Astronomy Program. We wondered about what kind of Earth changes we could tie to this scaly monster. Although, before we wade into the cosmology of the Crocodile and the Hippopotamus, let us become more familiar with the mythology. We can begin by reading what the great Egyptologist Wallis Budge stated in the *Legends of the Egyptian Gods*:

> *Rá proposed a sail on the Nile, but as soon as his enemies heard that he was coming, they changed themselves into crocodiles and hippopotami, so that they might be able to wreck his boat and devour him. As the boat of the god approached them they opened their jaws to crush it, but Horus and his followers came quickly on the scene, and defeated their purpose. The followers of Horus here mentioned are called in the text "Mesniu", i.e., "blacksmiths", or "workers in metal . . .*

The World Tree 191

Animal Cosmology on the Egyptian Burial Chamber of Seti I. (modified from original by W. Gaspar)

Figure 8-2: The Bull, Hawk, Vulture, Sun, Birth sign (Mos), Hathor, Hippopotamus, and the Crocodile are shown in Egypt. What is the exact cosmic role of the water tied Hippo / Crocodile duo? (Drawing by Wm. Gaspar)

—William Budge, Legends of the Egyptian Gods

The main players of the universal Mono-myth here again being brought into the scene by Rá, the Sun god, Horus, the Hawk-headed Prince, who connect the Sun to the Magnetic North, which stands for the Iron Core of the Earth. Thus, cosmologically the IRON Core of the Earth is connected to the followers of HORUS in the form of the 'blacksmiths or workers in metal'. This definitely supports our Cosmic Model theory. Now, we will obtain the help of the Hindu Rig Veda to see what happens if Virtra, the Dragon, is slain:

24: From the Rig Veda I will tell the heroic deeds of Indra, those which the Wielder of the Thunderbolt first accomplished. He slew the dragon and released the waters; he split open the bellies of the mountain . . . The Generous One took up the thunderbolts as his weapon and killed the first-born of dragons . . . Indra killed Virtra, the greater enemy, the shoulderless one, with his great and deadly thunderbolt. Like branches of a tree felled by an axe, the dragon lies prostrate upon the ground.

—Hindu Myths, Penguin Classics

Virtra, the Dragon, in Hindu mythology, is known as the Restrainer. This seems to apply to 'restraining the water' that the Crocodile Dragon is in control of. What was the state of the world before the water was released? Was Mother Earth dry and hot?

Did the biblical fiery Dragon fall by the famous British Knights to end the fiery heat? Was the release of the water a pleasant affair or was the dump of water just as bad as the dryness preceding it? The symbolical ferocious nature of the Crocodile / Dragon suggests to us that whatever they brought in the way of earth changes were not pleasant. Thus, the 'water broke' with the Crocodile somewhere during the week between April 23rd and May 1st. Did this watery disaster arrive after four months of heat and burning? Was this in the third, fourth, fifth or later years of the Last Twelve years of the Heroic Journey? We are determined to reestablish an exact sequence of events that will reveal to us the last twelve years of the old earth changes scenario and allows us to predict the last twelve years of what may be awaiting us after December 21, 2012.

In the Theban Judgment scene of the *Book of the Dead*, a belief is revealed that speaks of a beast that is part Crocodile, part Lion, and part Hippopotamus. This tri-partite Beast was called Am-mit ('*am*' means '*fire*' in Egyptian) that was there at the time when the souls to be judged were dragged in front of the mummified Pharaoh King. This pharaonic Judgment time is when the weighing of the heart against the feather is accomplished. The *feather* represents the *cosmic wind*, and the *heart* here may be the *heart of the Earth*. The gravitational based and seemingly non scientific question is whether; - could the cosmic or the solar wind blow hard enough to turn the Heart of the Earth? Naturally, the question is only poetic in nature, since we know that the 'Feather' won against the 'Heart' and the Core of the Earth shifted.

This truly childish appearing fairy tale system of feathers,

The World Tree

crocodiles, wild boars and hidden hearts preserved to us advanced cosmology. The combining of the creation legends from all around the world along with the Egyptian hieroglyphic pictures, the extensive Hindu mythology and the very accurate Mayan calendar allowed an intelligent search for the telling astronomy and mythology. From the very telling picture hieroglyphs of the ancient Egyptians to the Sumerian simplified cuneiform all the way to the Bible and the Koran, we almost lost most of the cosmological meanings.

The amazingly gifted Latino painter Raphael painted St. George and the Dragon in 1504. The act of eliminating the vicious Dragon is done mythologically to liberate the Libyan city of Silena from the Beast who requires annual human sacrifice. The story in itself does not contain any more hints of historical clues than one would detect in the classical Greek mythology, but it is still accepted as part of our

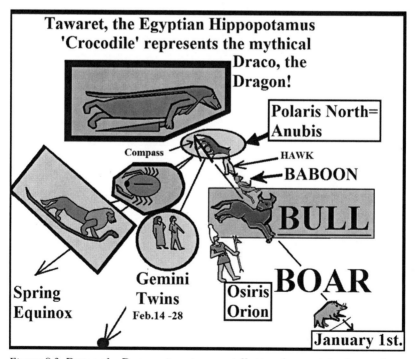

Figure 8-3: Draco, the Dragon, is a star constellation shows up in the northern sky from May 1st until about September 7th (Drawing by Wm. Gaspar)

Christian heritage. The killing of the Dragon by the Christian Saint George falls on around April 23rd. Then the astronomical dates of May 1st mark the appearance of the Crocodile / Hippo / Dragon brings to mind the Maypole Tree celebration that was usually done on the first day of May. Was there a great struggle in the past during the previous colossal earth changes when a hero had to fight a Red Dragon in the sky to stop the Fire and pour the Water? As far as the fiery dragon, we already have some ideas about what the sources are of the fire from the sky. Thus, the Fire was quenched by the god of the oceans. Then, darkness was on the land that was lifted by the birth of the Sun of the Son by the Ninth Hour. In the meantime, what was happening to Mother Earth and her closed legs, we do not have a detailed and very specific answer, yet.

As far as the Universal Mono-myth is concerned, it does not matter to us whether it was the most famous hero of Iceland, SIGURD, who slew the slimy Dragon, or if it happened to be Hercules and the Argonauts who achieved the deed; all we care about are the cosmological patterns that enlighten us to know how and when to expect the next tragic event.

Our classical Greek mythology tells the tale of Jason who, much like Hercules was educated by the wise centaur, Chiron, who trained the young heroes to be the most skillful and wisest pupils. Jason, the son of the dethroned king Aeson and his wife Alcimede, swore that he would punish his uncle, now King Pelias, who stole the Throne from his father and made them refugees from Thessaly. Thus, Jason prepared for his journey back to Iolcus, Thessaly, to punish his uncle or to at least die trying. In his long journey, he would go to find the **Golden Fleece of the magical RAM** (this is a good time to recall the Egyptian Amun) that was hanging from **an Old Tree guarded by a DRAGON!** We will tell the tale by H. A. Guerber's version.

Those of us who are cosmologically minded start remembering that one of the Old Trees of these old tales have to represent the Magnetic North and as we just learned, the Dragon constellation sits between the Two North Poles of the Earth's geo-dynamo from May to September on the Dendera Zodiac.

Jason set out to travel to Iolcus. It was in the spring when he reached a rushing stream that he almost could not pass through. An old woman was there who asked for help, and Jason, being a kind-hearted soul,

The World Tree

naturally helped her cross the perilous waters. In the rushing torrent, our prince lost one of his golden sandals.

Here we stop our tale for a short mythological reminiscing. In India, it is still a sacred tradition to wear only one sandal during the marriage ceremony. There is something similar in the Judeo-Christian teachings of the Bible, and possibly throughout the world. Who is the Hero of astronomy that has one FOOT in the RIVER? The answer is easy; it is ORION. His outstretched foot ends in the bright star RIGEL, a word that means *the foot* in Arabic. Now, that might not specific enough for some people, but that is why we encourage everybody to pick up those ancient and classical mythologies and creation legends – to shed a little light on (maybe illuminate a little light on) the old, enigmatic tales in view of our Cosmic Model.

Returning to the tales of Jason, we learn that, about the time he was ready to depart from the old Lady, she suddenly turned into a beautiful, young woman. The **PEACOCK** by her side made Jason to recognize her as Juno, the Queen of Heavens.

One might ask, in terms of cosmology or astronomy, what does the Peacock represent? It is an excellent question! It is positioned on the **SOUTH** side of the globe representing one of the southern Poles. Would we be surprised if this colorful avian species would show up at the 'Nativity' birth scene of Krishna and Christ? Well, we should not be surprised as the Peacock does show up as a further proof to the universality of cosmology from one corner of the Earth to the other.

Thus, our classical Greek myth continues. With the aid and protection promised by the Celestial Queen, Jason proceeded on to his uncle's city where the treacherous king was throwing a celebration in honor of the immortal gods. At the end of the ceremony, the arrogant king noticed Jason's naked foot missing a sandal. That is when he remembered the old prophecy he had been told, about being cautious around a man who came to him wearing one sandal. The ensuing confrontation between the Hero and the King sent Jason in search of the Golden Fleece of the Ram that was sacrificed to honor Helle. We learnt that Phryxus, the brother of Helle, placed the Golden Fleece on an Old Tree and he made a Dragon to stand guard twenty-four hours a day.

When Jason began his quest for the Golden Fleece, he properly asked for divine protection.

> *To secure Juno's assistance, he began by visiting her shrine at Dodona, where the oracle, a Speaking Oak, assured him of the goddess's goodwill and efficacious protection. Next, the Speaking Oak bade him cut off one of its own mighty limbs, and carve from it a figure-head for the swift-sailing vessel which Minerva, at Juno's request, would build for his use from pine-trees grown on Mount Pelion . . . When quite completed, Jason called his vessel the Argo (swift-sailing), and speedily collected a crew of heroes as brave as himself, among whom were Hercules, Castor, Pollux, Peleus, Admetus, Theseus, and Orpheus, who were all glad to undertake the perilous journey to lands unknown.*
> —H. A. Guerber, *The Myths of Greece and Rome*

Needless to say, that there are several heroes among the Argonauts who represent an important astronomical position in our Universal Legend! For example, the famous Gemini TWINS, named CASTOR and POLLUX of Jason and the Argonauts lead us back to the overheating of our globe. The Cosmic Lovemaking became marked amongst the Roman holidays where we celebrate the well known Valentine's Day. The numerous Greek gods and biblical elders seem to be repetitious about the same stories under different names to bring us the cosmological principles. Searching for a historical human hero would be a useless task to accomplish. One of the other famous Argonauts is named Theseus and in more than one important ways he repeats the deeds of the giant Hercules.

Then another Argonaut is the famous Orpheus who certainly does not need a lengthy introduction amongst the Greek heroes. Orpheus is the famous musician, the Harp Player of the Celestial Songs that cosmologically can only come from the Harp constellation that stands for the Magnetic North of the inner heat-producing geo-dynamo. These Greek astronomical heroes are unquestionably important players in a BIG OPERA of Cosmic proportions and it would take another lengthy book to detail their heroic tales. Since, we are only searching for commonalities amongst the world wide tales we encourage everybody interested in the subject to study up on these astronomical heroes.

One can also find references for giant heroes in the Chinese creation stories:

Origins, the Chinese say that in the very beginning nothing existed but formless chaos. Then some of the matter of this chaos collected to form a cosmic egg, in which the opposed principles of Yin and Yang were perfectly balanced. For an age the egg hung in the void, then it was quickened, no one is sure how, and within it took the form of Pangu, the First Man. He is generally portrayed as a 'Hairy Primitive Giant with an enormous Club, much like our own Hercules . . .
—Nigel Suckling, *Year of the Dragon, Legends and Lore*

Actually, we detect two primitive hunters of astronomy with a club or an axe in their hands, likely allowing an assimilation of two different heroes into one. One of the two club wielding heroes is certainly Hercules between the two north poles, but the other one is Orion, the Hunter, who defines the earth changes for us on the astronomical chart.

All these stories help us build an understanding that all over the world we celebrate the same astronomical holidays. For example, similar to the Egyptians, there are fertility ceremonies in India for the marriage of trees, the May tree celebration of Spring in Europe, the celebration of the Christmas tree in December for many cultures, the *Tree of Life* from the Judeo-Christian Bible, the legendary Scandinavian Ash tree, Yggdrasil from the Nordic mythology, and also the mythical tree with seven branches in Siberia. Interestingly the Slavs of Carinthia have a tree and a young man that they decorate on Saint George's day, generally on April 23, and after the ceremonies of song and dances, they throw both the young man and the tree into the water.

Now, let us search this date out to determine whether the Water and the Dragon have a relationship in the sky according to the Star Talker Astronomy Program and the alignment of the Zodiac of Dendera, which keeps a northerly look at the sky, with Polaris in its focus. We find a few expected surprises. First of all, the first stars of Big Dipper / Ursa Major appear in the sky on April 24 and it lasts until about May 7. Then the initial stars from the tail of the Dragon appear around April 29 to May 1 by the star called Giansar. Thereby, we were successful in tying astronomically the Christian Saint's days and religious holidays of the Egyptians, Jews and other great religions of the world to the universal Cosmology with actual dates.

St. George's Day is celebrated in Palestinia in the town of al Khader near Bethlehem.

April 23 is also the Day of Book, due to the fact that both William Shakespeare (by the Julian calendar) and the Spanish Miguel de Cervantes (by the Gregorian calendar) passed away on that date in 1616. Both poets were writing about parts of the Mono-myth in an enigmatic fashion.

It may only be an amazing coincidence that the rite of the Slavs of Carinthia was performed on Saint George's day, the killer of the Dragon. The fact that it ends in a watery disaster may hint at the relationship existing between the tipping of the Earth Axis, the Sacred Tree, and the Flood. Although there may be various other interpretations, it seems reasonable to expect that, when the Earth's Axis deviates from its position (when an axial shift occurs), then flooding, enormously large tsunamis and other disasters happen. However, when the Cosmic Tree becomes stable and stands tall, then we can celebrate the renewal of life, the creation of a new world, in essence. The symbolic tree itself is, therefore, the tool of the Creator to affect the renewal of all life forms after fires, earthquakes, hurricanes, darkness from volcanic eruptions, and floods subside.

We have not mentioned much about our Mayan Crocodile in the creation stories. In a southern Mayan town, there exists a stone monument that shows an interesting scenario. There are two trees, each with a bird sitting on top. The smaller tree is actually growing out of the body of a CROCODILE that is upside down. We are very sure about this Crocodile is the same concept as Draco, the DRAGON, since its body is shown tied to the sacred trees with a rope, even on the Maya stone carvings. Furthermore, when one studies the Mayan creation stories, the Popol Vuh, one can learn about ZIPACNA, the Crocodile, and how much trouble he is, not to mention that he likes to feast on CRABS (Cancer constellation) and the FISH that we are going to learn about in our next chapter.

On first peek at the above stela, it may not be obvious that the deity (Seven HunAphu), who is standing there holding the main tree, has his LEFT ARM totally cut off and is profusely bleeding. Knowing

The World Tree 199

that the cutting off of the arm can relate to an Axis shift, much like the mummified Egyptian pharaoh losing the use of his arms even when he is sitting on the Throne to Judge, we are reassured that this Mayan stela is part of the world heritage of Mono-myths.

Throughout the world, we encounter these animal symbols and the sacred marriage ceremonies connected to them. The Egyptians did not lack in festivities. They celebrated the Opening of the Mouth of the Crocodile, the beginning and the end of the Harvest, the Union of the Two Lands, also the Union with the Sun's Disk, the New Year, and the Divine Birth of the Goddess's child, and the celebration of the Cosmic Marriage that was held at EDFU and lasted fourteen days.

Two weeks prior to the new moon, the celebrations began in the town of DENDERA, the headquarters of the Goddess of the West, HATHOR. We learned that the third month of the Inundation, at the new moon, the festivities began to shift to the Nile, and the Goddess was placed into a boat to be transported upstream toward the south. The information on the months of inundation as far as the specific time of the year is still disputed. The lovemaking, drunken singing, and music lasted from new moon until the full moon of that same month. It was in a fashion probably not dissimilar to a Brazilian fiesta, the Greco-Roman Bacchus orgies for the celebration of the god of Wine, or the Mardi Gras of New Orleans. For the Sacred Cosmic Marriage celebration of HATHOR'S union, the beautiful Goddess of fertility and sexuality was shipped to Edfu, to the hometown of HORUS, where his temple was located. There, the two, a god and a goddess, spent two weeks together in "merry-making."

The concept of the male god at times was clearly represented on the walls of the pyramids with nakedness and an obvious phallic symbol (Min). In our secret of the Sacred Cosmic Marriage, it is MIN the Egyptian god with the erect phallus who will symbolically unite with our feminine VA. Thus, the word MIN-er-VA becomes decipherable. MIN, the Egyptian god of Fertility and Water will be mining the secrets of the VA of Mother Earth. Miner-Va this way becomes both the male and the female partner of this Cosmic Union. The Phallic symbol of the Egyptian MIN was not done for the pornographic shock value, or to represent the fertility of a coming harvest. It was mainly done as a reminder of that single day's event, which changed the lives of billions

Figure 8-4: The Phallic Symbol at Uxmal, Mexico in the Mayaland and also in Egypt (Photo credit Wm. Gaspar)

The World Tree 201

of people 11,700 year ago. That day, the male god, the Old Sun King, shed his seeds on Mother Earth. He showed his face and everybody trembled in fear.

This same male god, preparing for the Sacred Cosmic Marriage ceremony of the Egyptians and the Babylonians, is present in the case of the Mayans, as he is showing a Delta-shaped phallic symbol. It is amazing to realize that the Egyptians similarly have a DELTA shape of the mouth of the Nile pointing from the north to the south. Thus, examples of topography, river entrances, kingdoms and warriors from both countries point to a shared knowledge. This Mayan person with a delta shaped phallic symbol is truly revealing. Also, we noticed that that on this wall statue from the Mayan town of Uxmal—the headquarters of the famous Mayan kings, the enigmatic Xiu Family—we find this excited male deity with his head clearly separated from his body. We understand that some males lose their head when excited, but here we suspect that a different kind of cosmic beheading is implied. This again reminds us that the Old Sun King does get overheated with love, especially by Valentine's Day, but the time of his beheading is near.

This phallic symbol is a very telling and exact sign that can lead us onto the right path where this enigmatic shape of sacred geometry can reveal itself. It is not a coincidence that the DELTA shape is chosen for the male god. 'Del' in the Romance languages usually indicates some form of 'possession' and the 'TA' we already explored. Thus, the concept of the 'Del-Ta' also stands for the Cosmic Union.

We assume that the clever and well-educated, cosmological minded astronomy priests of any great nation would not have put up derogatory, sexually explicit material unless, in their Sacred Cosmic Marriage ceremony of the spheres, it was required. We think that their controversial material was very carefully selected to maintain a well developed and sophisticated system that, on the surface was vulgar and childish but, at the core, hid a very complex set of astronomically inspired material. This simple and shocking way was not as shocking for the future generations as the telling the brutal truth about what happens to the Earth when we are in need to charge up its dynamo every 12,000 years.

The more we know about how simplified they made a potentially very complex system, the more we stand in awe of these ancient priests, artists, astronomers, and linguists. We could probably attach a few other titles to the names of these scribes, but we have not even discovered all of what they were trying to say and portray.

In the amazing Maya cosmic art in Uxmal, one can find the exotic deity displaying the Delta sign as the phallic symbol. For most people, it would not mean much, but we understand that there is a deeper cosmological depiction beyond the simplicity of the representation. When one can detect the same 'delta' male force concept from the Ancients from different continents, then it behooves us to admit that a universal mono-myth had to exist for thousands of years. The male force in this picture is not referring to an elevating human experience, but rather the Magnetic North Axis force due to the Sun king eruption will violate the receptive Polar North of Mother Earth and unify with it. This cosmic erotic love play between the gods is definitely a major disruption to our globe's harmonic existence, but as we know by now it is a necessary event in our solar system and for the Earth's geo-dynamo to be recharged. If these forces would not follow the commands of the nature god, then life could not exist on our planet with our periodically failing magnetic field that needs to be recharged. The upside down Pentagram is the secret of this cosmic sex and it is also the secret of the Phoenician and Roman letter making. The concept is called the Squaring of the Circle.

This is as much we are willing to reveal on that at this time. Interested seekers should research the topic on their own.

Therefore, we understand that the phallic symbol represents a gigantic male eruption and the subsequent axis shift that destroyed civilizations in a blink of an eye, and probably earned a male designation to the genderless Creator Spirit. This is what most our major cultures celebrate in their festivities.

Staying with the Slavic cultures, as well in Siberia and Japan, the Bear carries a lot of significance in the renewal ceremonies. The star POLARIS represents the North Pole star, and we know this because this particular star does not seems to move like the other stars in our galaxy, and that is because it is almost straight up over the North Pole, and from the northern hemisphere, the North Star looks like it stays in one place. It has been roughly in that position for the last thousand years, with the astronomical sign of the Bear (URSA MINOR, or the Little Dipper) in the polar region, it is not to difficult to perceive how the ancients might have blamed the Little Bear in the sky for moving the polar axis out of its place. In this scenario, via animal and astrological symbols, many cultures today remember the axial shift of the Earth and the following renewal of life through the sacrificing or celebration of the Bear.

The Sacred Bear of the southwest Indian tribes, combined with the lightning symbol, carries the same message: When the axis of the Earth changed into its new position and pointed at Polaris, the Little Bear was perceived to give rise to a lightning bolt of celestial fire at a magnitude never seen before. Without hesitation, we honestly are inclined to believe that all ancient religious stories, mythologies, and legends directly or indirectly recall these astrological figures as part of the fellowship of the Magical Tree with the intention of preserving the sacred cosmic knowledge of the timing of Noah's flood and the astronomical basis of the related natural cycle.

We all know the first written version of the Deluge in which the *Sacred Tree* has a major importance. From the library of King Ashurbanipal, at the biblical Nineveh in Assyria, comes the most ancient account of the Flood. Written on clay tablets, *The Epic of Gilgamesh* has been translated by Maureen Gallery Kovacs, among others. It is an old Babylonian legend, which predates Abraham's biblical story of the Flood. Historians and religious scholars are well aware of the fact that this epic or a related common source is what provided Abraham, the founder of the Judeo-Christianity, with the flood legend. A true biblical scholar needs to investigate the Babylonian, Egyptian, and Mayan legends, along with the stories of Abraham and Moses to discover a common pattern in the similar cosmological wisdom. This ancient astronomical knowledge we deciphered for those who might think that a sudden shift will occur in Nature soon. Those, who although may thrive on high octane adrenalin rush, still may prefer the added security of meaningful earthly preparations. The task is not easy, the challenges are almost insurmountable

It is clear to us that Gilgamesh is the astrological figure of the Perseus constellation, Humbaba is Taurus, the Bull, and Enkidu, the Hunter, is the Orion star constellation with the Cedar Tree of Lebanon being the Axis of the Earth in our theory. We all know by now that

Gilgamesh's quest to kill the Bull of Heaven appears very similar to the bull-killing scene from Mithraic religion. Enkidu, who represents the astrological hunter, Orion, is missing from the Mithraic scene. In

Figure 8-5: *The Phallic Symbol at Uxmal, Mexico in the Mayaland and also in Egypt (Photo credit Wm. Gaspar)*

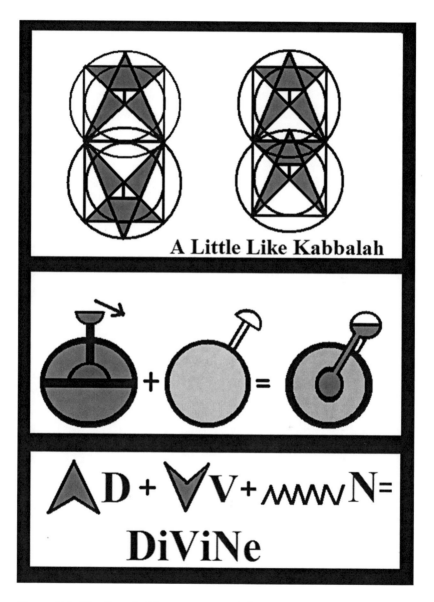

Figure 8-6: The Cosmic Union concept of the Male and Female Drawing by Wm. Gaspar)

other sections of the Epic, the hero, Gilgamesh, is equated with the Bull himself; he is the young prince who is credited with bringing the earth changes.

Humbaba, the Bull, is pictured with horns, but has a human body. Could the slaying of the beast, sphinx, bull, etc., be what will cause the darkness? The seven circles in the sky may represent the seven sisters of Pleiades, who play a significant role in the creation legends of numerous native tribes; the two Sioux Indian warriors who attack the White Buffalo Maiden could also parallel this image. This is the imagery, the same symbols, and astrological time frame that the Mayans and the biblical book of Revelation talk about.

We invite our readers and scholars to read again the *Epic of Gilgamesh*, because it has been mistakenly read as a poetic telling of a hero's quest for immortality, or simply as a historical event in a king's reign. On reading this old tale, one unavoidably notices the pieces of the core mono-myth, which is definitely astronomical in nature. These stories were written by the old Babylonian bards thousands of years ago, with some variations over time. The *Epic of Gilgamesh* was more likely rewritten and recopied over the centuries or even millennia, prior to Abraham's story of Noah's Flood. The wandering nomadic Jews preserved these Babylonian tales, and later incorporated them into the current biblical wisdom.

The Ainu people of Japan, from the island of Hokkaido, are shown performing the Bear Ceremony at their New Year celebration. The astronomical reasons behind the power of the bear, the dragon, and the harp as a musical instrument are beginning to provide clarity to the students of ancient enigmas about the World Tree. The Bear cub sits on top of the mythological world tree and spins the Earth around its axis.

In Nordic and Native American mythologies, the Bear is an important cosmological reminder for the Polar North. The Hawk flies by the Dragon to get to the Bear or the Throne. We had learned that the axis shift traveled through the entire length of the Dragon constellation in the sky. The use of the Dragon symbol by Saint George, the Chinese, and the Greeks appropriately illuminates the universality of this Beast as a threat against humanity. It makes perfect sense to slay the Dragon since it is trying to destroy our Mother Earth. Now we know why John of Patmos reveals to us through his book of Revelation that the "fiery red dragon" plays an important part in his message. The ancient saintly prophets knew that Dragons are a dangerous threat to the survival of humanity, and we should know this, too since one of them is growing in power for us just around the corner.

In his recent work on Chinese Dragons, Nigel Suckling writes the following lines:

> *Dragons in China are revered as basically wise and gracious divine beings, great benefactors of humanity because they are masters of the weather; natural disasters are attributed to their carelessness.*
> —Nigel Suckling, Legends and Lore

Further on his book, *Legends and Lore*, Suckling states about Chinese astrology: "They also commonly employ the shoulder blade of cattle." And in our theory, the shoulder blade of the Bull is the Pleiades, thus ancient Chinese astrology and legends preserve this same wisdom. We hope that by now we have shared our theory in enough detail that the reader already has an understanding of how all these ancient culture wrote down the story of the renewal of life through the Cosmic World Tree.

Interestingly, the Shoulder Blade of the Bull was mentioned in Chinese astrology in the last quote from Mr. Nigel Suckling. This is such a small world. The Bull in our theory is the astrological sign of Taurus and the Seven Sisters' star cluster (the Pleiades), is positioned on its shoulder. Special importance is given to the position of the Pleiades in the timing of the astrological clock of the ancients. It clearly represents darkness, power over the light, and the long awaited return of the Sun.

Thus, creation stories dwell upon a re-creation at 11,700-year intervals. Let us quote from Professor Ulansey:

> *The MITHRAS liturgy is a section from one of the Greek magical papyri which announces itself as being a Revelation granted by (the great God Helios MITHRAS). At one point in the magical ritual described by the next seven Gods appear who are called the "Pole*

Lords of Heaven," and who are greeted as follows: "Hail, O guardians of the pivot, O sacred and brave youths, who turn at one command the revolving axis of the vault of heaven." Immediately after the appearance of these seven pole-lords another God appears, "a God immensely great, having a bright appearance, youthful, golden-haired, with a white tunic and a golden crown and trousers." This God continues the text, is (holding in his right hand a golden shoulder of young bull: this is the Bear which moves and turns heaven around, moving upward and downward in accordance with the hour).
—David Ulansey, *The Origin of the Mithraic Mysteries*

There are several interesting connotations to be derived from this passage. One is that the seven gods, the pole lords of heaven likely represent the Seven Star Brothers in Ursa Minor, the little Bear. This Little Bear whose tail is tipped with Polaris currently hovers directly above the Earth's north Pole. The Seven Brothers are cleverly tied to the Seven Sisters of the easterly Pleiades from where the Sun erupts from. The Pleiades is represented by the hidden meaning of the shoulder blade of the bull. Then the youthful, golden-haired hero, who is a Christ-like figure, also fits the description of MITHRA / PERSEUS / GILGAMESH. Furthermore, the hero just as easily could be the Mayan Kukulcan or the Aztec hero, Quetzalcoatl. These Native heroes, whose principal symbol was the Feathered Serpent, used very similar ideologies to represent the birth of the Son of the Sun. Almost identical legends are attached to other South and Central American deities of different tribes, where the hero is called Gucumatz, Votan, Itzamna, or the Peruvian Viracocha.

Thus, the Mithraic mysteries represent a proto-type for all heroes all around the world. The direction of the Seven Sisters of Pleiades connects the action to the north celestial pole, and signals the time when sunlight disappears. All this could begin to happen for us again in our times around the years 2012 and 2013. Therefore, **the birth of the divine Child is tied to the opening of the gates of heave**n and the axis shift scenario. Let us quote from the Bible:

Now learn the lesson from the fig tree: As soon as its twigs get

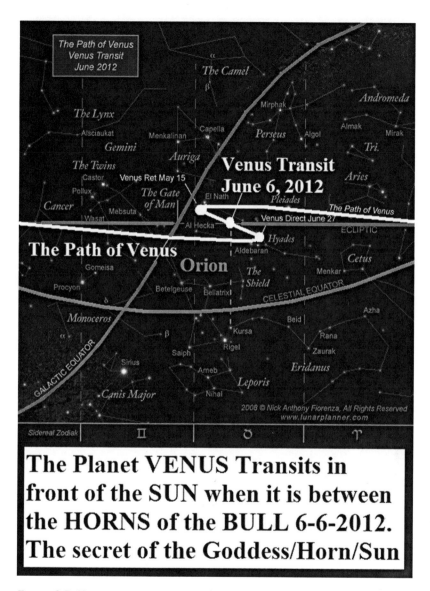

Figure 8-7: Here we can see the Crossroad at the point when the Sun in the ecliptic crosses the Galactic Equator. Also, we can see the Seven Sisters, the Pleiades constellation to be above the Taurus constellation as the shoulder blade. (By José Jaramillo & Nick Fiorenza)

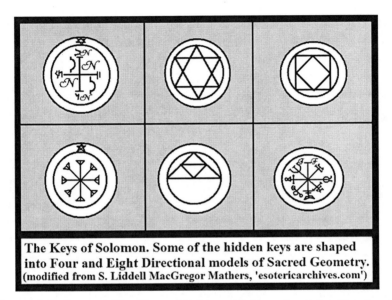

Figure 8-8: Some of The Keys of Solomon display the eight directional models we are following (Drawing by Wm. Gaspar)

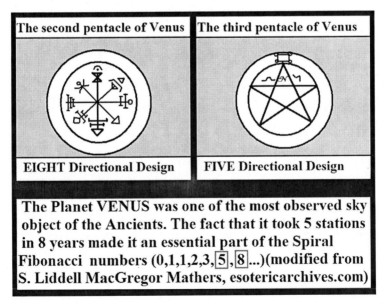

Figure 8-9: The Planets in the Key of Solomon (Drawing by Wm. Gaspar)

The World Tree

tender and its leaves come out, you know that summer is near. Even so, when you see all these things, you know that it is near right at the door. I tell you the truth; this generation will certainly not pass away until all these things have happened.
—Bible, Matthew 24:32-34

Let us see our next figure to find out more truth about the World Tree and the Bull of Taurus constellation, and his relationship with Pleiades as the shoulder blade of the Bull. Definitely, all these cosmic players are an important part of our theory and of the fellowship of the Magical Tree, also to identify what could be the significance of the term door that could be a place in the sky that is known as "the gate of men" near GEMINI constellation where the Sun crosses the Galactic plane.

In actuality, the force that arrives from the Black Hole to the Sun and subsequently to Earth is an unseen electro-magnetic super-wave. It will cause a great illumination from the Sun when his SON is born. That ancient depiction of a fiery star-like phenomenon with a smoky tail approaching Earth could have been a meteor to warn us, but we are certain that it was the enormous solar flare that caused the burning of the entire globe in the past. These events happen rather unexpectedly, out of its season, and are disruptive to the idyllic harmony of cyclically predictable nature.

The last few graphs of this chapter will introduce the symbols and divisions connected to the Keys of Solomon.

Just examining the previous two graphs for the basic building blocks in the Keys of Solomon right away one can notice that there are shapes of the Pyramid, the Squares in the Circle, the Four and the Eight divisions all lead us back to our Pentagram and Delta symbols for the Unity of Male and Female Forces at Zero Time. There are no separate and specific meanings for the different sounding mysteries of the Egyptians, the Babylonians or the Judeo –Christians and not even the Nomads or any Aboriginal tribal cultures. The only meaningful mysteries about the earth changes is the timing and the manner of the Magnetic Axis shift to the Polar North every 12,000 years and its subsequent slow resolution over decades.

CHAPTER 9

The Great Fish WHALE

8: For everyone who asks receives, and he who seeks finds, and to him who knocks it will be opened.
9: Or what man is there among you who, if his son asks for bread, will give him a stone?
10: Or if he asks for a fish, will he give him a serpent?
—Bible, Matthew 7:8-10

This work would not be complete without revealing the cosmological secret of the biblical Whale and Mayan Great Fish. The ancients had to employ different views to their Three Dimensional model of the Milky Way Galaxy to explain certain phenomenon with respect to what is happens during these dramatic, but natural Earth changes. Animal tales of all kinds were utilized. Some of them we are already familiar with, but the Whale who swallowed Jonah, in the Bible story, remained unexplained, so far. The ancient astronomy priests were trying to emphasize major power sources in our Milky Way Galaxy, such as the THREE PLANES! Those three Planes or Equators are the:

The World Tree   213

1. GALACTIC PLANE
2. SOLAR PLANE
3. EARTH'S EQUATOR

Many times, we pay too much attention to the Axis of the Earth or the Sun and the Galaxy. Sure, those are the parts of the Sacred Tree, but we rarely speak of the Planes of those systems. Actually, in some ways, the knowledge of the Plane of the Galaxy is much more important than the wisdom of its axis for our purposes.

As children, we all saw Earth globe models in our classrooms, and may have even wondered why the Earth was shown tilted on its Polar North. That is a long explanation, but at least we have seen it. The

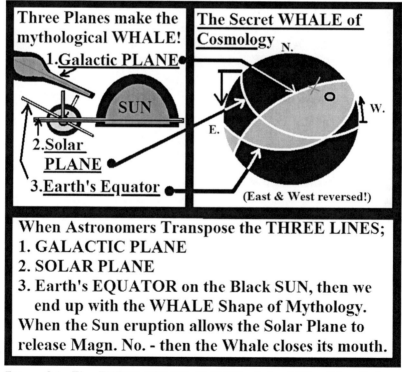

Figure 9-1: The THREE POWER PLANES of the Milky Way Galaxy. (Drawing by Wm. Gaspar)

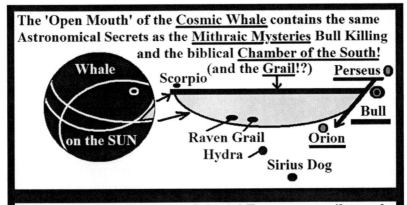

Figure 9-2: The Mouth of the Whale is the same as the Chamber of the SOUTH, and is the same as the Mithraic Bull Killing; that 'Mouth' is the difference between the Equatorial Plane and the Solar Plane (Drawing by Wm. Gaspar)

Equator, which is the PLANE of our Earth, is then also tilted at about a 23.4 degree angle from the horizontal of the Solar Plane. Thus, both the Mouth of the Whale and the Mouth of the Crocodile are opened 23.4 degrees, leaving the evil number of 66.6 degree remnants from the 90 degree square.

From the studious research of Professor David Ulansey, we have this mythological fact certain. His book, *The Origin of the Mithraic Mysteries*, laid the foundations for similar research. The Hero, Perseus / Mithra, the Bull, Dog, Snake, Raven, Grail, and Scorpio all line up perfectly on the Celestial Plane.

Above, when the Axis shifts, then the Whale blows out water; that means the biblical subterranean fountains erupted from the pressure of the turning Core of the Earth. The last few years a new phenomenon, the bright colored blue electrifying noctilucent clouds

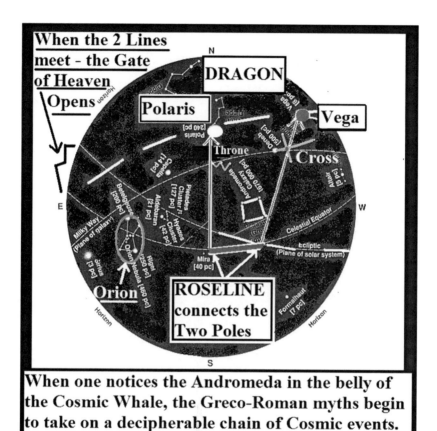

Figure 9-3: The Trinity Power Lines, when transposed on the Sun, will make the shape of the Great Fish or WHALE of cosmology.}

attacked our skies that resemble a Blue Great Whale in the Sky. If it happened similarly in the past, likely years before the Cataclysm, than it was a great symbolism to have the Great Blue Whale tied to the swimming catastrophe. This huge Cosmic Whale is coming toward us again on the resonating living water of the Galactic and Solar planes. Examples of the noctilucent clouds resembling great blue whales can be found on the exciting web page ran by NASA and NOAA and titled the 'Spaceweather.com'. Those Astronomy Priests from the sunken continent of Atlantis, who assembled the original scientific tales to withstand the test of times were not only geniuses, but they had to be raised in an environment where scientific understanding of the workings of the Galactic and Solar Plane along with the Two Pillars of the Earth were readily understood. We have not arrived to that general understanding, yet.

Thus, this watery symbolism was maintained by most cultures and tribes. Poseidon, Neptune or the Native American Water Monster whose babies were snatched away, thus it came to the people to reclaim them. The more widely read classical Greek mythologies present us with their share of Watery Monsters from the Sea. The Greek gods and monsters are not satisfied anymore with little babies, they are demanding a

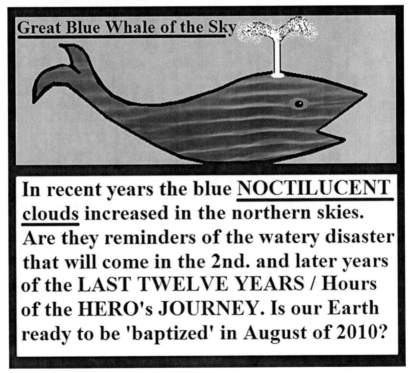

Figure 9-4: The noctilucent electrifying clouds in the north resembling a Great Blue Whale flowing in the living water of the sky. Is this one of the signs the biblical fathers mentioned that we would only see toward the End Days\—Modified by Wm. Gaspar from 'Koeman', from Spaceweather.com}

The World Tree 217

beautiful Virgin every year. Thus, we learn about the love stories of the Beast and the Beauty over and over again, whether it is an 800 lb. Gorilla called King Kong or it is a slimy Loch Ness like monster.

The tales of watery monsters are countless, but in our cosmic explanation for the role of the Whale we are strictly searching for one astronomical and cosmological point. We do not see another way to explain the magical tales of the classical Greeks or the biblical Whale of Jonah.

> *8: And it shall be, in the day of the LORD'S sacrifice, that I will punish the princes and the king's children, and all such as are clothed with foreign apparel.*
>
> *9: In the same day I will punish all those who leap over the threshold,*
>
> *10: "And there shall be on that day," says the LORD, "The sound of a mournful cry from the Fifth Gate, a wailing from the Second Quarter, and a loud crashing from the hills.*
>
> —Bible, Zephaniah 1:8-10)

In the above quote from the Bible, we hear that there will be punishment to the king's children. The cry will be from the Fifth Gate. By now we probably do not have to repeat the obvious hidden wisdom that is becoming everybody's celestial knowledge. The Fifth Gate is the Fifth Hour of the Egyptian Book of the Gates that has twelve gates. In the fourth gate, there is fire, and in the FIFTH gate, there is the picture of the Lake of Fire that refers to the oceanic changes during this severe Judgment. The Fifth Hour of the Last Twelve is when we shall witness the earth changes on a grandiose level involving the axis shift and the ensuing upheavals of the oceans. When was '0' Hour prior to the beginning of the Last Twelve Hours according to the sacred Venus Calendar of the Mayans?

We are certain, that something as important as counting the last twelve years before the huge Flood, the astronomy priests had to chose a marker that was very rare, very telling and something that was constantly studied in the sky. Our candidate for that special date is the planet Venus when it was in the Center of the Sun on June 9th,

2008. Thus, by the summer of 2010 we are in our Second Hour of the Last Twelve Hours. Therefore, soon we should witness the validity to the assumption that the Fourth Hour will bring us the Sun eruption resulting in the Lake of Fire and the Axis shift in the Fifth Hour.

The few years left to find out whether we are correct or not are not a lot of time. It is not a whole lot of time to reconsider to actually physically prepare for a possible disaster or not, but we still think it is enough. There are safe places and willing groups to take on more members who are serious about preparing for what might be coming. One of the most important and easiest aspects of preparation is shopping for food. We have not seen any pair of healthy females, who could not drive into the parking lot of a Sam's Club with a U-Haul truck and not be able to drive away a few hours later with $ 10,000 worth of canned and boxed groceries. There is certainly much more to the idea of the urgent preparation, but the first step has to be the acceptance of the idea that there is a high chance for enormous earth changes to occur. When that is materialized in the mind, then it has to be deeply rooted in the spirit. Out of the tens of thousands of people, who might intellectually accepts the idea of these coming earth changes, there may be only a handful of seekers who will make meaningful preparations according to the excepted magnitude. With our work, we hope to increase the number of believers and preparers.

As we continue to make our case for an urgent preparation, on our the next graph we would like to observe the Pharaoh who is sitting on his throne in the middle of the Judgment. There are ten people standing in line that we translate as ten years of earth changes. The first one has a scale on his back, and there are nine standing in line, each one step below the other. Is this the basic proto-type concept for the nine levels of the Ladder of Jacob?

In the Solar Barque, there is another Baboon who is shown as a beheaded animal, may be the sacrificial Lamb, since the Baboon lines up around Christmas in the sky. Most astronomy legends portray Perseus / Mithra as a magician wearing the Phyrigian Hat. In the Dendera Zodiac, there is no Nordic Brother or there is no Perseus or Mithra, but only the Baboon. Baboon is human-like, but not exactly, and they are known to represent loud screaming, sudden fear and anger. Thus, when the star Algol in the star constellation Perseus will align around December 25th of 2012 or 2013, will we witness something that will

The World Tree

create a sudden fear in us? Further with the Baboons, we commonly remember them as athletic creatures jumping between trees. The Baboon is also present at the Nance Tree of the Mayan creation story and in the Hindu legends of the Two Brothers.

The famous Mira star of CETUS, the Whale, is inviting the seeker to look at it as a miracle. The Mouth of the Crocodile and the Whale explains why the Egyptian RÁ priests decided to spell the name of the

One of the Egyptian Judgment Scenes. Pharaoh sits on the Throne (Cassiopeia) The NINE levels refer to Nine Years. The Baboon in the Solar Bark cuts off the Head of the BOAR. The Year here is not clearly marked with the Four Bulls.

Figure 9-5: There are 9 gods to be judged and the 10th is holding the scale on his shoulder. In the upper right-hand corner, we have Anubis, the Jackal, standing for the Polaris North, and below him is the Baboon who represents Perseus / Mithra (Drawing by Wm. Gaspar)

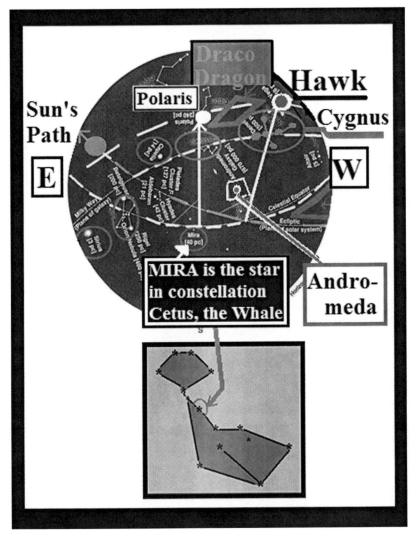

Figure 9-6: The Whale in Astronomy (Drawing by Wm. Gaspar)

RÁ Sun King with a MOUTH (R) and an ARM (Á) when talking about the Sun.

It would have been easier to depict the Sun as a yellow disk with rays, but there were the secrets of the mouths and arms the ancients needed to incorporate into the name of the Sun King for us to rediscover

The World Tree 221

this enigma with the right key and affirmations. The Jewish Torah, the Old Testament of the Bible, does caution us not to pray to Baal (Bál). They did not think that we were going to pray to a Whale; rather, they wanted to keep the conversation going about this deeply-held secret in their holy books without revealing the hidden cosmological meanings. They were just using the usual tactic: keep everything enigmatic in the plain sight of the people then they will not search for it. Certainly, given any treasure written down and locked away in a forgotten cellar, someone will eventually dig to find and discover that hidden secret. Similarly, too, given so many other religious and mythological secrets, it would be a grave mistake to search for an actual grail or historical figure in the sacred tales of advanced cosmic knowledge. Religious leaders and their followers throughout the ages have propagated this mistake. This was the way to maintain very important information about the timing and the mechanism of the severe Earth changes that happened at the beginning of the Age of Leo about 12,000 years ago. Certainly, it is likely that we will witness the occurrence of similar Earth axis shift at the end of the Mayan calendar in 2012 or 2013. According to some scientists, it may not happen until the Mega Sunspot cycle of 2040, but we would not bet on that. Thus, let us forget about the old trickery of transforming the divine into human, as that is only the tale that carries the sacred message of unavoidable periodic Earth destruction. Let us now reverse the trend and make the human tales the divine message of our creator.

We are certain that neither the talented Dan Brown nor his extremely clever editor believed the marvelous ending they manufactured for their fiction story. They only applied the old trick of **humanizing the divine** to be able to deliver some important secrets to the public that would unlock the first three doors of the last seven on a corridor of cosmic knowledge. This hallway would be leading up to the unwrapping and opening of the Pandora's Box of secrets.

In this astronomical set-up, we can observe a tilted column, a belt such as Orion's, a column, a crab, and the Serpent. This makes us believe that the origin of the Mayan and the Egyptian cosmology sprang from the same root, a long time ago. They might as well be brothers in the legends of the Atlantis. Thus, both the Mayans and the Egyptians place us back to the Cancer Constellation on the Ecliptic

Figure 9-7: It is NO wonder that the Mayans, too, have a Great Fish on top of the Head of this Eagle Warrior. Observing the Eagle Warrior of the Maya, it becomes obvious that the Tilted Column, the CRAB cutting off the Head of Hydra, perfectly fits our Cosmic Model. Permission from Scala / Art Resources, New York}

The World Tree 223

where the BEEHIVE is located. Actually, in the Mayan pyramid town of Tulum, located on a Mayan beach of beautiful white sand with extensive coral reefs, one can find the main figure of the Sacred Child's birth on the walls of the temples. These centrally-located pictures and murals on the walls depict a baby emerging toward the ground, and are connected to the Beehive.

This place of the birth of the upside-down, descending god is above the gate which is commonly depicted with a letter T shape. That is the center of the Lip of the Whale or the place of the Cancer star constellation, it is the Red Lobster on the dinner table of the Greek hero Theseus, and it is also the Donkey present at the Nativity scenes. This is how complex and interwoven Astro-mythology can be.

These diving gods wore wings; according to the experts and they represented the Bees.

We are not surprised, and we realize that the Universal Mono-

Figure 9-8: The Cancer star constellation near the Beehive star cluster is the Center of the LIP OF THE WHALE. It is also the place where all previous 'gods' were born, including the Descending Bee God of the Mayans (Photo courtesy of William Gaspar.)

Figure 9-9: The Descending God of Mayan Tulum. The Birth is accomplished in the Gate between the Two Five Decorations. In the symbolic language of the Mayan Astronomer Priests this point represents M-44, the Beehive Cluster where the other Sons of the Sun were born in different mythologies throughout the World. (Photo courtesy of Wm. Gaspar)

The World Tree

myth is in full throttle in the Mayan mythology. It is no wonder, then, that the cover page of *The Lost Tomb of Viracocha*, written by the excellent researcher Maurice Cotterell, contains the Peruvian CRAB depicted with a human face. The astronomical Cancer / Crab are the Crib of a number of Sons of the Suns, since the Mono-myth is always the same.

Rather than being a physical object, it is clearly an astronomical GRAIL (Crater) in the center of South Celestial Equator (Chamber of the South) that shifted and turned the ocean waters to blood color by the erupting underwater volcanism.

This Sacred Feminine is the vessel into which the Equator of the Magnetic North (Solar Plane extension) will join. The Egyptian letter 'K' represents the 'Holy Grail' that is depicted by the Cup that is the lower half of the globe. Thus, in essence the Lower Lip of the biblical Whale would also double up for the larger 'rim of the Grail' that will contain the cosmic blood after the water turns to blood.

When that upper lip of the Whale closes down on the lower lip, the

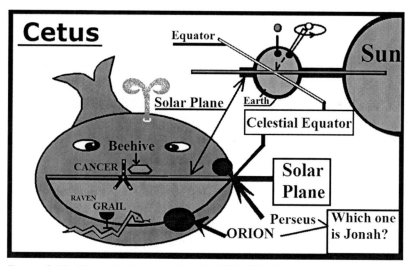

Figure 9-10: The Holy Grail is the SACRED FEMININE, except it is NOT A PERSON; rather it is in the middle part of the South Celestial Equator toward which is the Axis Shift progressed in the past. It is in the Center of the Lower Lip of the Whale. (Drawing by Wm. Gaspar)

Figure 9-11: In the Egyptian hieroglyphic spelling, the Town of the Crocodile, that is CROCODILOPOLIS, is the same as Vulva, the place of the Divine Birth Canal of Axis shift (Drawing by Wm Gaspar)

TWO Egyptian or biblical KINGDOMS UNITE! The constellation in the sky that holds together the two hemispheres, the North and the South, is the famous Hunter Orion / Nimrod.

One can surely hear of a few more secrets related to these astronomical and cosmic enigmas, but it has to be the individual seeker who should study the sacred writings, along with astronomy, sacred geometry, linguistics, and a number of the other sacred sciences. Without the self-sacrifice and rigorous study of these subjects, one cannot climb the ladder of Illumination. If one is provided with every

The World Tree

little detail, one will not be allowed to own the completeness of the sacred knowledge. The ancient cosmic priesthood was slaughtered or quietly died out a long time ago. Maybe they are still among us, but difficult to recognize. Now that we are not sure who knows what, and to what extent, who will pick up the sacred, royal scepter of cosmic teachings? Who will be brave enough to labor under the expectations of the future generations to carry this astronomical knowledge over to the new ages? Carry the ancient cosmology knowledge, we must, or the quickly approaching next catastrophe might wipe out the majority of the humans and, with them, the traces and methods of tools for meaningful reconstruction of the Mono-myth. Where are you hiding, you reincarnated Cosmic Priests of Rá?

The Grail is at the bottom of the Cosmic 'Birth Canal' that is open and suddenly closes when it gives Birth to the Son of the Sun. The COMPASS shape of the TWO NORTHS symbolizes the mouth of the Crocodile / Dragon and the Vulva.

It is no coincidence that the spelling of the Vulva and the Town of the Crocodile is almost identical in the Egyptian hieroglyphic language, since cosmologically – and only cosmologically! – do these two concepts have their origin together. That is an additional proof that we have truly found the secret formula of the ages, which connects otherwise unrelated ideas, animals and humans in a Form that only makes sense in a twisted appearing Cosmic Knowledge.

CHAPTER **10**

The Lost Codices of Pakal

> Thirteen Heavens of decreasing choice, nine Hells of increasing doom. And the *"Tree of Life"* shall blossom with a fruit never before known in the creation.
> —Lord of the Dawn

> But first we need to tell you about the boy-king of Mexico. He too, was born through an immaculate conception 1,405 years ago in the jungle of Mexico. He taught his people the super science of the sun, worshipped the sun as the god of fertility, and performed miracles. When he died they say he went to the stars. They called him the feathered snake.
> —Maurice Cotterell, *The Tutankhamun Prophecies*

The Aztecs left to us **the prophecy of thirteen heavens and nine hells**. Not a lot of people know what this prophecy is all about. It is important to understand our past to foretell the future. Before we go further into an explanation of this ancient prophecy of divine time, we need to go back to the past and understand where this future telling was coming from. In order for us to understand this Mayan-Aztec Prophecy of the thirteen heavens and nine hells, we visit

The World Tree 229

the time of the classical Maya. We need to especially study the ancient site of Palenque, a place that, at one time, was called NAA CHAN, the House of the Serpent. Some others called this house XBALANQUE. This mystical place was the home of one of the greatest Mayan kings who ever made history for the Mayan civilization. His name was Lord Kinich Janaab Pakal II; everybody referred to him as **Lord Pakal**. Whether mythological or historical, **Lord Pakal was born on March 23 of the year 603,** and he was the son of Kan Bahlum Mo and Queen Sak Kuk. **He was ceremonially proclaimed king of Palenque on the exact day of July 29 of the year 615 at the age of 12**. He received his crown as the **eleventh king of Palenque** from his mother, Queen Sak Kuk.

Here is the story of Lord Pakal:

At the age of 29 on June 8, 632, Lord Pakal had an inspiration in the spiritual teaching of Tezcatlipoca. That was same year of the birth of his first son. On this date, June 8, 632, Lord Pakal decided to be an inspiration for others; also, he knew that was the moment of preparation, not only for him, but for the mission to lead his people in a spiritual journey. Two years later, his second son was born, and four years later, his third son was born.

In the year 645, Lord Pakal sent his three sons, Hun Nahb Naah Kan Tiwo'ol Chan Mat, Kan Bahlum II, and Kan Joy Chitam II to Teotihuacan to learn everything from Pakal's master, Tezcatlipoca. They spent thirteen years studying the four books of prophecy. When they had learned all the secrets, they returned to Palenque. Years later, following the murder of Tezcatlipoca at Teotihuacan, the four books of prophecy became the texts of his people. The books were guarded in Teotihuacan at the Temple of the Jaguar, where they were kept in secrecy. Then the four books of prophecy were moved to the sacred city that had been built and dedicated to Tezcatlipoca. This city was in the jungle lowlands; its name was XBALANQUE (Small Swift Jaguar), known today as Palenque, There at Palenque, the four books of prophecies were studied and tested. A number of copies were made of them, so more students could share in the wisdom found within their folds.

On the year 663, a great Lord came to Xbalanque. No one knows where he came from. He was an unusual man among the Mayans. This lord or priest was a very strong man, and very wise. He became a very good friend of Lord Pakal. His name was Ajawmuni. After spending some time together learning from each other everything about astronomy, cosmology, and spirituality, and the secret behind the Mayan count of days, he became an equal of Lord Pakal in the teachings of Tezcatlipoca. By following the examples established by the Priest Ajawmuni and Lord Pakal, together they solved many social and religious problems for their people.

The sacred books of the Mayans were then hidden for future generations. The enigmas contained in those books were about the mysteries of the calendar cycles of creation and the endless harmony of time related to the World Tree.

The books compiled by Tezcatlipoca were written on deerskin, and Lord Pakal feared they would be destroyed by wear if a more lasting surface were not found for the sacred words. At last, and by way of a Mixtex artist, a suitable surface was brought by Ajawmuni. He was a master of the lost wax process of casting gold, so he taught Lord Pakal, who learned a way to pour melted gold into sheets, paper thin and capable of lasting forever. On 22 sheets of gold, Lord Pakal, helped by his friend Ajawmuni and his three sons, copied the sacred prophecies that were laid down by Tezcatlipoca. Carefully, they duplicated every glyph, every astronomical correlation. These twenty-two sheets of gold represented the prophecies of the Fifth and Sixth World of 13 Heavens and 9 Hells. When the tablets were finished, Lord Pakal sent them back to the Temple of the Jaguar at Teotihuacan, and he kept the originals and the copies. "Soon the people from Teotihuacan will be in great need of these books," said Lord Pakal. And so it was done in that way.

In the year 680, Lord Pakal had lost his first son, Hun Nahb Naah Kan Tiwo'ol Chan Mat. This was the sign for him that his time was close to an end. So he gave his last test to his two sons and to his loyal friend Ajawmuni, to protect the wisdom that soon would be needed for future generations. Their mission was to engrave in each stone of

The World Tree

the magnificent city the ancient cosmological knowledge, so they could remember it in the future. Lord Pakal died on August 31 of the year 683, at the age of eighty, having totally committed his entire life to his Lord Tezcatlipoca. His accomplishments as a holy man were a testament to a phenomenal spiritual life. During the last 36 years of his life, he erected

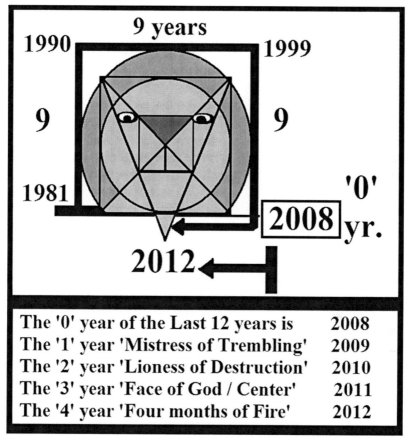

Figure 10-1: Showing the last 36-year long path of the Sun during the alignment with the Center of the Galaxy, known by the Mayans as XIBALBA Be. Notice that the last 36 years started in the year 1981, reaching the center in the year 1998-1999, and getting into the season finale by the year 2016. The Fourth Step from the bottom 2012 is only a half-sized step. Uxmal, Mexico (Photo by William Gaspar.)

the "forgotten temple," which was built to honor the memory of his parents. Then he built the "Dark House" mentioned by Bishop Nunez de la Vega. Within that the Dark House, he placed the history of his four past worlds and the instructions he had tested during his lifetime, for a world of great peace. He engraved in each stone the stories about the last shift of the ages taking place here on Earth.

Whether Lord Pakal's life was real or enhanced by mythology, he still left important cosmological information of the renewal of life on Earth, by adding the cosmic knowledge into his own historical story. He was one of the greatest kings of Mayan civilization. The last 36 years of his life, he erected the Forgotten Temple.

The 36 years are the metaphor for the last 36 year-long path of the Sun in his ecliptic through the Galactic Center. As we detailed earlier, it takes 72 years for the fixed stars of the sky to move ONE DEGREE of PRECESSION. Since there were Axis Shifts every 12,000 years, the 36 represented the HALF. Also, since the Sun appears about the size of a half degree of Precession, according to John Major Jenkins, we assume that this measurement was incorporated into the Sacred Cosmic Knowledge. Our friend and fellow author, Don Antonio Gaspar Xiu, a direct descendant of the Xiu family of Mayan kings whose family built the Uxmal complex, beginning with 751, brought our attention to the cosmologically-minded construction.

Let us see our next figure from Uxmal, Mexico:

This astronomical and cosmological marvel of the Mayans was built by the Xiu family of Astronomy Priests over 1,200 years ago. The building is divided into two levels to mark the Heart of the Galaxy at the position of the 18th step half way through the 36 steps representing the solstice Sun's Half Degree Precessional travel and alignment with the west. In this simple temple from Uxmal, one can learn about the structure of the last 36 years that is then end in the Last Twelve steps corresponding with the Last Twelve Years of the Hero's Journey. Amazing, is not it? How can any theory, but the existence of a multi cultural Atlantis explain the findings in several different cultures the identical representation of the Last Twelve Years of the Hero's Journey on a 36 year base? One cannot easily bring into this picture a common ancestry for the mythologies of the Last Twelve Years that are found in such diverse ancient cultures as the Hebrews, Egyptians, Greeks,

The World Tree

Figure 10-2: The 2012 Pentagram and the Thirty six year counting (Drawing by Wm. Gaspar)

Romans, Chinese, Mayans and the English. Unless, there was a common and accepted knowledge base of Cosmology as we rediscover now, and a common glorious past either right before or after the Flood, it would be difficult to explain the almost identical recounting of a past tragedy with such accuracy.

We can observe which year the tip of our Pentagram is pointing in this 36 year long journey. It is the YEAR 2012!

The symbols of the Mayans and the symbols and legends of the newly arrived Spaniards hid the same secrets, but at that time the differences seemed larger than the similarities.

The story of Lord Pakal is phenomenal. Like so many other seers of the future, he was able to tell the time of his own death. On the beautiful morning of August 11, of the year 683, exactly twenty days before his death, he called his two sons and his loyal friend, Ajawmuni, and he told them the vision that he'd had that morning.

It will be the time when foreigners will bring their own gods to convert our people!
—Lord Pakal prophesized to his people.

By history, we know that Hernan Cortez arrived in the Maya lands to bring Christianity. If it was not for a few Catholic priests and bishops, and scholarly-minded Spanish conquistadors who may have realized that there was important Sacred Knowledge of the ages contained in the Mayan teachings, then the rest would have burned all the books of the ancients. Luckily, a few manuscripts remained that missed the pyromaniac Spanish Inquisitors. These manuscripts, including the Dresden Codex, the Madrid Codex, and a few others, allow us to retrieve some of the Mayan beliefs, and realize the obsession and the accuracy these astronomers had about following the path of the planet Venus. In the codices, there were accurate tables about the rising and setting times of the planet Venus. Important symbolic nature gods were attached to some of the dates, to remember the significance of particular reappearances of Venus. Without that knowledge—simply relying on other cultures for the same information—we would have been in a more difficult situation.

The stories are definitely there, in a hidden format, in other cultures religious beliefs and creation legends, but the very accurate

Figure 10-3: Mayan and Egyptian signs seem to match. (Photo by Wm. Gaspar)

astronomical precision that the Mayan astronomer priests maintained for this age would have been lost if the conquerors had been successful in burning everything. Luckily, instead of eliminating the entire Mayan astronomical knowledge, the leaders of Cortez's troops realized that they were dealing with a superb culture with important knowledge for everybody on the face of the Earth. This important cosmology was

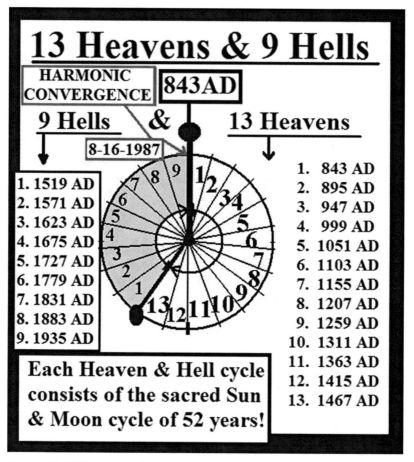

Figure 10-4: We can see here the 13 cycles of heavens each 52 years, also the 9 cycles of hells each 52 years. This fifth prophetic round ended on August 16, 1987, with the Harmonic Convergence Prophecy (Chart by Jaramillo and Gaspar.)

saved by the actions of the Mayans, mainly, but there were Spanish and European scholars who greatly contributed to the modern understanding of those codices. We, at times, feel separate from others or we feel different from nature, but we are not, because we all are one and the Cosmic Knowledge is ONE!

We all know that each of the 13 Heavens were made of 52 years and each of the 9 Hells was made of 52 years cycles. Quetzalcoatl was born in the year 947. These prophetic rounds of hells ended with the Harmonic Convergence Prophecy in August of 1987. The interesting thing that we notice here by following a prophetic cycles of divine time is that the very next day, on August 16, 1987, another prophetic round was opened, but this time each 22 cycles (13 Heavens and 9 Hells) were only made of 360 days. Counting in the Gregorian calendar 13 years from 1987 would take us to the year 2000, and 9 more years would lead us to 2009. Would anything happen in the year 2009? We hope that Christmas of 2009 will not mark the beginning of larger signs of what to expect soon. We would like to believe that we have time to prepare for December of 2012.

Let us continue Lord Pakal's story:

After these 22 cycles each of 52 years, then another prophetic round of 13 Heavens and 9 Hells will be open, but this time each cycle will last only 360 days for the total of 7,920 days, then it will come the time of 40 days and 40 nights for the time redemption, you will be tested, and then after all of this again for another 1,260 days, then you will need to redeem yourself for another period of only 60 days, if you keep yourself faithful to the divine time, then you will have the opportunity to see the end of the cycle. Learn from my knowledge brothers and sisters, find the truth behind the voice of my spirit, because I am Pakal son of Kan Bahlum Mo and Queen Sak Kuk, I proclaim myself to be the messenger of the time cycles, and remember that all my numbers are secret for those who can understand it. Once that I had gone you will see clear signs and you will be tested, so learn of my knowledge and be one with the eternal light that lies within us.

The World Tree

The story of the saintly life of Lord Pakal was carried far and wide by his people. As always happens to true culture heroes, the story of his life became fused with the histories and cosmology of other culture heroes. The legend became fused with the histories of Tezcatlipoca and Quetzalcoatl and others, until, at the time of the conquest, it was indeed very difficult to find the true history of Lord Pakal.

Explanation of the 13 Heavens and 9 Hells Prophecies:

The fact that Lord Pakal was 29 years old, we think, could hide something in the cosmological dates of the 36 years of the path of the Winter Solstice Sun in front of the Galactic Heart? The 29th year of the 36 year path of 1980-2016 would be again 2008 or 2009. We know that June 9 of 2008 was the magical date when the planet Venus was found in the dead center of the Sun. That sacred count of Venus only happens every 121.5 years, thus it may be that is what the reference is about.

It is difficult to ascertain, but we definitely feel that these ages of the Mayan kings had a prophetic message. Also, we would like to point out that another interesting event was occurring in the other side of the world; this was also the day when the Prophet Mohammed ascended into the Heavens. The June 8 date is very close to the date JUNE 9, 2008, when the planet VENUS was in the Center of the Sun in the West in the middle of the last pair of Venus Transits (The Portal) of 2004–2012!

This might be a very interesting fact in the religious history of humanity, but certainly could be an important day marker in the cycles of cosmology. Returning to the first figure of this chapter, one can observe the outlines, sizes, and measurements of the steps in Uxmal, Mexico, to see that the Xiu family tried to leave us the sacred divisions in those 36 steps!

When Lord Pakal built his first temple, it was on January 7 of the year 646, which in the sacred count of days was 12 OC, which means 12 Dog kin 90. His first building was the Forgiven Temple, built in honor of the memory of his mother, Queen Sak Kuk and his father, Kan Bahlum Mo. This temple became a sacred place for Pakal because it was the place where he honored the living memory of his ancestors.

Then, on August 10 of the year 663 a great lord came to Palenque.

His name was Ajawmuni. This day was 1 IX, which means 1 Jaguar kin 14. And on August 31 of the year 683, Lord Pakal ascended to the heavens, and in the sacred count of days, that was 6 ETZNAB, which means 6 Flint or Mirror kin 58. He gave to his son, Kan Bahlum II, and his loyal friend, Lord Ajawmuni, the task of saving the cosmic knowledge for the future generations.

For twenty years, Kan Bahlum II followed his father's orders and he became a very important king of Palenque. He constructed several structures during those years, such as the Temple of the Cross, the Temple of Sun and Moon and also the Foliated Cross Temple. He dedicated nine years of his life to completely sealing the tomb of his father, and he was responsible for putting the 620 inscriptions in the central panel of the Temple of the Inscriptions. This was erected to honor the two great cycles of the Mayans, which are the 360 days and the 260 days cycles.

The dedication of the tomb was the year 692. Following that, Kan Bahlum II ascended into the heavens in the year 703. Lord Pakal had told his son that after the twenty years of the recollection of the sacred cosmic knowledge, it would be the time that came to be known as the Seven Lost Generations. Each of these seven generations lasted twenty years, making 140 years. So this prophecy was from the year 703 to the year 843, when the prophecy of the 13 Heavens and 9 Hells started on September 2 of the year 843. This is the sacred count of days named 1 IMIX, which means 1 Crocodile kin 1. This day sign is the beginning of the sacred count of the Mayans.

After 13 Heavens, each of 52 years, we arrive at the end of the 13 cycles of heavens. This period ended at a precise moment in history, on the day of April 21 of the year 1519. This was the time when the conquistador Hernan Cortez landed in Veracruz and that was the beginning of the 9 Hells.

That day in the sacred count of days was 4 OC, which means 4 Dog kin 30. This sacred sign is one of the 16 sacred signs around the lid of Lord Pakal's sarcophagus.

Was the arrival of the Spaniards foretold or was it coincidental to that date, which fact may have saved their lives and allowed the newcomers to completely take over the Mayan Empire.

Did Lord Pakal know the exact moment when the conquistadors

The World Tree

would land on their shores?

Did this incredible, spiritual king of the Mayans traveled in Spirit ahead of his time to see the future? Was he the son of a god, to possess the supernatural ability of foretelling the future, or were all those dates only lucky coincidences for the advantage of the conquistadors? Nobody is sure about that, but one thing is for sure—we need to look at his prophecies carefully and attempt to understand the message of counts and dates.

When Hernan Cortez landed in Vera Cruz, it was the beginning of the 9 Hells, each hell lasting 52 years, until the time of August 15, 1987: the time of the fabled Harmonic Convergence.

This specific Harmonic Convergence was the end of these 9 hells. That day in the sacred count of days was the day 13 SUN, which means 13 SUN kin 260.

A new prophetic round started on the very next day of the sacred calendar, on August 16, 1987. On the day of 1 Crocodile kin 1, a new TZOLKIN round was opened, with a new prophetic round of another 22 cycles of divine time. That day was the beginning of a new cycle of hope for the people of the Earth that the prophecy of the 13 Heavens and 9 Hells was fulfilled.

The Prophecy of the 10 Kings of Mexico:

In 1952, the Mexican archeologist, Alberto Ruz, discovered the Pakal's exactly 1,260 years later. The prophecy was received in Mexico and was sent to all principal authorities. It was also clear that the prophecy speaks about the 11 Serpent, and this sacred sign refers to one of the 16 signs around the edge of the lid of Lord Pakal's tomb.

The prophecy says:

The false king will return to be embraced by the 11 Serpent, do not doubt any of this. When the moment comes for him who is to be the eleventh in succession he will be the 11 Serpent then join the believers and will begin the prophecy that liberates from all kings and successors ... Ten witnesses I place around my tomb, Ten Kings were in my time before I was proclaimed King of Naa Chan ... I am the eleventh.

Here is a list of the eleven kings of TENOCHTITLAN:

1. ACAMAPICHTLI: from 1375 to 1396
2. HUITZILCHUITLI, son of ACAMAPICHTLI: from 1396 to 1417
3. CHIMALPOPOCA, son of HUITZILCHUITLI: from 1417 to 1426
4. ITZCOATL, son of ACAMAPICHTLI: from 1427 to 1440
5. MOCTEZUMA I ILHUICAMIN: from 1440 to 1468
6. AXAYACATL: from 1468 to 1480
7. TIZOC: from 1480 to 1486
8. AHUITZOTL. brother of TIZOC: from 1486 to 1502
9. MOCTEZUMA XOCOYOTZIN: son of AXAYACATL: from 1502 to 1519
10. CUITLAHUAC. brother of MOCTEZUMA: last 80 days of 1519 A
11. CUAUHTEMOC of ANAHUAC, son of AHUITZOTL: from 1519 to 1520 {End NL}

While NAA CHAN, now known as Palenque, flourished, there were ten kings of Palenque before the eleventh in succession, who was Lord Pakal. We find similarities between the kings of TENOCHTITLAN and the kings of NAA CHAN.

Here is the list of the eleven kings of Palenque:

1. KUK BALAM: from 431 to 435
2. CH'AWAY: from 435 to 487
3. BUT'Z AJ SAKCHIC: from 487 to 501
4. AHKAL MO-NAHB I: from 501 to 524
5. KAN JOY CHITAM: I from 524 to 565
6. AHKAL MO-NAHB II: from 565 to 572
7. KAN BAHLUM MO I: from 572 to 583

The World Tree

8. NAA KANAL IKAL: from 583 to 604
9. AAK KAN: from 605 to 612
10. SAK KUK mother of Pakal: from 612 to 615
11. PAKAL II from 615 to 683 {End NL}

Around the lid of Pakal's sarcophagus, we can find the ten emissaries, which are the ten kings in succession to the throne of Palenque before the eleventh, who was Pakal. The transition at the end of the 13 Heavens dated from June 14, 1999 to June 8, 2000, exactly four years before the first Venus Transit of 2004.

Here is the chart of each cycle of heaven of modern time:

Heaven number 1: Started: August 16, 1987
                              Ended: August 9, 1988   9 Sun

Heaven number 2: Started: August 10, 1988
                              Ended: August 4, 1989   5 Sun

Heaven number 3: Started: August 5, 1989
                              Ended: July 30, 1990   1 Sun

Heaven number 4: Started: July 31, 1990
                              Ended: July 25, 1991   10 Sun

Heaven number 5: Started: July 26, 1991
                              Ended: July 19, 1992   6 Sun
Heaven number 6: Started July 20, 1992
                              Ended: July 14, 1993   2 Sun

Heaven number 7: Started July 15, 1993
                              Ended: July 9, 1994   11 Sun

Heaven number 8: Started July 10, 1994
                              Ended: July 4, 1995   7 Sun

Heaven number 9: Started July 5, 1995
                              Ended: June 28, 1996   3 Sun

Heaven number 10: Started June 29, 1996
                 Ended: June 23, 1997   12 Sun

Heaven number 11: Started June 24, 1997
                 Ended: June 18, 1998   8 Sun

Heaven number 12: Started June 19, 1998
                 Ended: June 13, 1999   4 Sun

Heaven number 13: Started June 14, 1999
                 Ended: June 7, 2000   13 Sun

Here is the chart of each cycle of Hell:

Hell number 1: Started June 8, 2000
              Ended: June 2, 2001   9 Sun

Hell number 2: Started June 3, 2001
              Ended: May 28, 2002   5 Sun

Hell number 3: Started May 29, 2002
              Ended: May 23, 2003   1 Sun
Hell number 4: Started May 24, 2003
              Ended: May 17, 2004   10 Sun

Hell number 5: Started May 18, 2004
              Ended: May 12, 2005   6 Sun

Hell number 6: Started May 13, 2005
              Ended: May 7, 2006   2 Sun

Hell number 7: Started May 8, 2006
              Ended: May 2, 2007   11Sun

Hell number 8: Started May 3, 2007
              Ended: April 26, 2008   7 Sun

The World Tree   243

Hell number 9: Started April 27, 2008
                 Ends: April 21, 2009   3 Sun

Human history goes from August 16, 1987 to June 7, 2000 just like it did in the past—from the time of King Pakal that lasted from 843 to 1519.

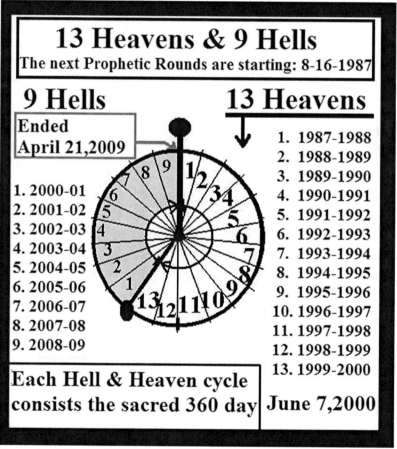

Figure 10-5: Here we have the new prophetic round of 13 heavens and 9 hells each of 360 days; the end will be on April 21, 2009, exactly 490 years after the Spaniards landed in Mexico on April 21, 1519 (Drawing by Gaspar & Jaramillo)

One can also observe that the nine cycles of Hells in this prophetic round lasts from June 8, 2000 to April 21, 2009, and will end exactly on the anniversary of 490 years after the conquest of Cortez on the day of 3 AJAW kin 120, which means 3 Sun Kin 120. Then from April 22, 2009, we open a new prophetic round, but this time it is known as the 40 days and the 40 nights, for a total of 80 days of preparation, just as it was with the 80 days of Cuitlahuac as the King of TECHNOTITLAN. On the exact day of April 22, 2009, which in the sacred counts of days is 4 IMIX, which means 4 Crocodile kin 121 is the beginning of the 40 days and 40 nights round of preparation that lasts only 80 days. So by April 21, 2009, is the end of the nine hells in the last prophetic round, and a new prophetic round of hope opens for humanity. Now, we begin the discovery how everything works in our beautiful galaxy within our solar system and our Mother Earth. It is since May 1980 that our Sun began his journey on the ecliptic and his path through the Galactic Center that lasts 36 years, until May, 2016.

Here is the chart of 40 days and 40 nights:

First 20 days: Will start: April 22, 2009
                                           Will End: May 11, 2009   10 Sun

Second 20 days: Will start May 12, 2009
                                           Will End: May 31, 2009   4 Sun

First 20 nights: Will start June 1, 2009
                                           Will End: June 20, 2009   11 Sun

Second 20 nights: Will start June 21, 2009
                                           Will End: July 10, 2009   5 Sun

On July 10, 2009, 5 AJAW (Ahau), which means 5 Sun, we started a new period in which we all need to walk together as brothers and sisters. On this path, we are thought of as beautiful flowers in the Garden of God, each one of us with different colors and different beliefs, but all staring anxiously at the sky.

This time, we will be observing the next round—the time of 1,320 days. This may also be a metaphor of the 13:20 of the Sacred

The World Tree

T'ZOLKIN. For 1,320 days, we will be walking this Earth knowing that the end of the cycle is near, and in the precise moment according to our calculations and our own *World Tree theory*, February 21, 2013, will be the beginning of the next Great Mayan Cycle on the exact day of 1 Wind kin 222, then it will be the return of the King.

A special day happens on February 22 of the year 2012. The Mayan Elders from Guatemala will celebrate their New Year. It will be the day 13 Earth of the Year 13 Earth. Another interesting fact is that the current Mayan Year, which started on February 22, 2008, was called the Year of the 9 Earth. This day sign is guided by the day sign 9 Reed. It is interesting to observe that July 4, 1776, when declaration of Independence of the United State of America happened, was the day 9 Reed kin 113, and in Mayan Long Count notation was: 12.8.0.1.13:

And it was written as follow:

12 BAKTUN x 144,000 = 1,728,000 days

8 KATUN x 7,200 = 57,600 days

0 TUNS x 360 = 0

1 UINAL x 20 = 20 days

13 KIN x 1 = 13 days

Days from the beginning of the Cycle =
1,785,633 days (Day 9 Reed)

Did Pope John Paul II fulfill the Catholic prophecy of the popes?

More interesting facts in human history, if we decide to use the Mayan calendar to understand the Catholic prophecy about their popes, are the dates related to John Paul II. It is really interesting to find out that May 18, 1920, when he was born, in the Mayan calendar was the day of 10 Sun kin 140, and that day was a solar eclipse in Europe, so it was clear to some observers that the prophecy was talking about that John Paul II, as he would be known as the Pope of the Sun and the

Pope of the People.

So when John Paul II was born there was a Solar Eclipse, and maybe a part of the prophecy was fulfilled. Interestingly, in the Mayan calendar day signs, he was born on the day of the 10 Sun, also.

When this well-liked pope, John Paul II, passed away, on April 2, 2005, that day in the Mayan calendar was the day sign of 5 Sun kin 200. For some believers, the prophecy was fulfilled, not only because he was born under the day sign of 10 Sun, but because he died under the day sign of 5 Sun.

All of these Mayan Day sign calculations could be just a pile of coincidences, because the numbers and the sacred signs always remind us to an important part of cosmology. Anyway, these dates mean something to a lot of people.

The Seven Years of Prophecies clearly states that the return of the Children of the Sun will occur at the end of the Mayan Cycle. In some form, one can even interpret this to the Old Testament prophecies, since IS means GOD in the Sanskrit languages and RA is definitely means the SUN GOD in Egypt. Therefore, the Children of Is-Ra-El may be the cosmic part of this ancient prophecy. These seven years of prophecies started in the year 1993, at the exact moment of August 11, 1993, when Pope John Paul II went to Mexico to IZAMAL to ask to all the Mayan Elders of Mexico for forgiveness for everything that the Catholic Church had done in the past. We think that the apology came a little too late, but was definitely well-deserved! Dr. Jaramillo was there in that precise moment in time to witness the speech that the pope gave in front of all the Mayan Elders. This was a great moment in history, when two lost brothers met to unite their spiritual and cosmological knowledge, rather than think of annihilation of each other.

Mass killing of Jews, Native Americans, Africans, or Nordic Barbarians would never achieve the intended goal of unification under the old Roman flag, since all those cultures and spiritual beliefs have come back much stronger in the last few decades. Learning from each other and finding the commonality in our spiritual and cosmological messages can be the only vehicle that will unite us as One, one of the Ones with shared knowledge and responsibilities for this beautiful Mother Earth! Only respect breeds mutual respect. We have most of our pieces to the puzzles, so let's place them on the round table of

Figure 10-6: Don Gaspar Antonio Xiu is the descendant of the famous Mayan Astronomy Kings who built Uxmal beginning around 751.

cosmology and make our knowledge One!

At this point, we would like to agree with Don Gaspar Antonio Xiu, who is a great friend and also the last descendant of the Great Mayan astronomy kings.

Don Gaspar Antonio Xiu is a living descendant of the Mayan royal family of Uxmal. Beginning in the year 751 the famous Mayan astronomy kings of the Xiu Dynasty carved the most amazing cosmological wisdom on the walls of the buildings and pyramids. The current Don Gaspar Antonio's fame of royalty reached the British Empire, and even Queen Elizabeth II visited this soft-spoken king of the Maya. By ancient Mayan laws, he is the current Mayan king of Uxmal, and he carries some of the Mayan prophecies related to the end of the Mayan calendar cycle. His message to the world is to live in a peaceful coexistence, to respect the rights of others, and to love one another without bearing grudges and hatred, in order to prepare ourselves for the world of tomorrow. That New World will arrive by the will of our Creator around 2012, and certainly not by human design. This is what we learn from the Mayan prophecies. Hatred, power supremacy, human exploitation, envy, and grudges must end. Humans are not the product of civilized power, but of the spiritual forces of the Creator.

The last time the three of us were together, in May of 2008, in San Diego, where all three of us were speakers in a 2012 Conference, Gaspar Antonio Xiu told us that the Mayan culture flourished in a time where there was no interference from any foreign culture or foreign power. Now we have a mixture of ancient cultures meeting in Mexico, and we need to cherish that. The interference from foreign powers into aboriginal communities is the source for hatred and discord, resulting in human life entangled in wars. This will not change unless we begin to learn from each other and recognize the commonalities rather than the differences. We are in this now together.

Also, he told us that, for him to be the royal descendants of the last Xiu kings to govern Uxmal and Mani means learning responsibility for others, not a power-oriented responsibility, but a responsibility of guidance to the world. He told us that the Mayan people are a kind and peaceful community, a culture that loves art, science, and astronomy. The Mayan worldview is peaceful and studious. The responsibility that he has as the last descendants of the Mayan kings is to preserve cosmic wisdom and, among all things, peace and human coexistence. He told us that humanity has to fight to end the threat of famine, and to stop wars that are the products of men who flaunt financial power. Money and power are the symptoms of the destruction that approaches. Mayans are spiritual people who want nothing to do with the thirst for supremacy, financial ambition, and the power of man over man. Mayans only would like to build spiritual power in these trying times.

He also related to us the history of centuries of generations of their Xiu lineage. Today, the great patriarchs of his family have disappeared. His father and his grandfather have passed away. Now he is the oldest of this direct lineage of the XIU family of Mayan astronomy kings. Interestingly, the combination name of GASPAR ANTONIO is a very meaningful set of names from their royal ancestry and the dynastic power. They are the recipients of this lineage and it cannot disappear over the centuries.

We appreciated Don Gaspar Antonio's sincerity and wisdom.

Our own Dr. Gaspar from Hungary can also relate the Gaspar name to the biblical Three Kings. The Magi had among them a King Gaspar, who visited Jesus at his Nativity Birth. The Gaspar name in some form goes back to the Persian and Jewish names where Gaspar or Kaspar meant the Treasurer, mythologically, and it was also one of the representations of the THREE STARS in the BELT of ORION! Then we also know about a Madam Gaspar in France who prophesized. Even further back, to the Hindus, the original Hindu name was KasiApa, which apparently meant The Tortoise, and in that KasiApa form, King Gaspar, the First King of the White People, mythologically, was the founder of the capital, Colombo, in Sri Lanka.

Now, we also know that there was a prince named Maya, both among the Egyptians and the Hindus, and naturally, the Maya survived with the Mayans. Our own Dr. Jaramillo's daughter is named Maya, and she also reminds us that these ancient names once were universal. At one point, we were truly only ONE great NATION.

The Spanish conquistadors brought the Egyptian, European, Judeo-Christian teachings to the Mayans to realize that we are all brothers from an older time, when all of this cosmic knowledge was ONE. Our spiritual job now is to bring the pieces of that once-great, unified astronomy of the ages back to everybody and to rejoice in the fact that we all say the same spiritual facts in a different dialect. That interconnectedness of "brother meets lost brother" on the other side of the globe is a larger spiritual achievement for us than just merely surviving some possible earth change scenarios.

This is our combined message: Unite and learn from each other. Learn only the common cosmic wisdoms that will unite us. Observe the basic Cosmic Wisdom in all the teachings from the past, as we start realizing that it is from the same ancient root that we all have originated. We all seem to possess an important segment of the whole, but nobody knows it all alone anymore. Now, our spiritual calling is to bring part of this old metaphysical wisdom into an international melting pot, and from the rusting scrap metals and heavy lead, let us attempt to synthesize pure metaphysical gold for the future. These cosmic wisdoms will not fade away with passing times and changing fashions, since these natural rhythms are the results of our Creator and not one human or one race will be able to wipe out the work of the Creator that lasted billions of years!

CHAPTER **11**

Decoding the Tomb of Lord Pakal

[Y]ou who speak with the Heart of Sky and Earth, may all of you together give strength to the reading I have undertaken."
—Popol Vuh

On June 15, 1952, Mexican archeologist Alberto Ruz discovered the tomb of the most important king ever in the history of Mesoamerica. Three years before he discovered the tomb, when he was in the chamber on the top of the Temple of the Inscriptions, he noticed there was a hollow tube coming out of the roof of the temple, and that hollow tube ran from the roof to the walls and all the way to the floor of the chamber. After he carefully cleaned the area, he noticed four pairs of circular holes in one of the flooring slabs at the top of the temple. After this, he was able to lift the slab clear of the floor to expose the limestone steps leading down all the way to the bottom of the staircase.

He spent three years excavating with his team, from 1949 to 1952, until he reached the bottom of the staircase. He noticed that the hollow tube that ran all the way from the top of the temple to the bottom of the staircase entered another chamber that had a triangular door.

They removed that door and, on June 15, 1952, Alberto Ruz discovered the tomb of Lord Pakal. This was after exactly 1,260 years of exile since Pakal's son, Kan Bahlum II, had sealed the tomb of his father in the year 692. Since its discovery it became known as the "Earth Spirit Speaking Tube," from where Pakal was communicating.

We want to point it out that after Pakal's death on August 31, 683, which in the sacred count of the Mayans was the day 6 ETZNAB Kin 58, which means 6 Flint or Mirror, his son, Kan Bahlum II, spent nine years sealing the staircase that led into the tomb. He sealed the staircase layer by layer. The day in the long count that was called 9.13.0.0.0 is when the dedication of the Temple of the Inscriptions occurred. The final dedication was in the year 692. Thus, from that moment on, to the time when Alberto Ruz discovered the tomb, exactly 1,260 years passed. Is it a coincidence or spiritual design?

When we went to the Temple of the Inscription in Palenque, we noticed that, in order to enter into the tomb of Lord Pakal, we needed to go down first 26 steps, and then we landed on a big step, and then we had to turn to the right to another flight of steps. Then we needed to go 22 more steps to be able to get to the place where the chamber is. In the interior of that chamber is the place where the sarcophagus of Lord Pakal is resting.

We noticed that it added up to a total of 48 steps to get all the way from the top to the bottom where the chamber was placed. Thus, in order to go all the way into the chamber and then come back to the top of the temple, one needs to walk through 96 steps. Walking one way down on 48 steps symbolically added up to the 48 pairs of Venus Transits, each 121.5 years apart from the next pair, marking a Quarter Precession of 5,832 years. The 96 steps symbolically represented to us the Venus Transit Pairs in front of the Sun that happens 96 times in an actual half cycle Precession of 11,664 years. Certainly, adding 72 years of temple building to that would sum up to an 11,736 year cycle, which would generally average out to be 11,700 years. That is an approximation of what we consider a Half Cycle Precession of the Equinoxes.

Furthermore, on the other side of the world, if one wants to count the number of the Two Ladies (Venus Transit Pairs!) hieroglyphic symbols on the tomb of King Tut in Egypt, we will end up with 24 signs on each of the 4 sides; that will again point to the 96 pairs of Venus Transits in front of the Sun.

We would like to add that there were also another four more steps

The World Tree 253

to enter into the sacred chamber that made the steps count to be a total of 104. That number equals the number of years for the Sun, Moon, and VENUS to meet up in the same point in the sky. Thus, we had TWO important Venus numbers—96 and 104—incorporated in the steps leading to the mythological Lord Pakal's resting place. Furthermore, the average of two Venus numbers 96 + 104 = 200 / 2 = 100 would yield the number 100, which also carries an important measure in mythology as the number of leagues between the two pillars of the Earth. Also, the soldier who killed Jesus Christ was a Centurion; that word means something like a One hundred in Latin, further giving us the feeling that any of these holy men had sacred cosmic calendar rounds incorporated into their lives, which seem to be mythologically constructed.

One interesting connection, according to our theory, is that takes about 96 pairs of Venus Transits (96 x 121.5 = 11,664 years) to complete a half cycle of Precession of the Equinoxes to arrive to the shift of the ages. The lid of Lord Pakal's sarcophagus contains sacred signs that only one who is prepared can read it. On the west side of the lid are nine sacred signs, and on the east side are another nine sacred signs, for a total of eighteen sacred signs. The necklace of Lord Pakal was made of 130 beads. Our surprise was this: When we multiply these two numbers (130 x 18), we get the special number that has something to do with Venus cycle of 2,340 Venus years. In the Dresden Codex, according to Adrian Gilbert and Maurice Cotterell, we find a special number that relates to the Venus Cycles.

That number is 1,366,560 days. In solar years, that would be approximately 3744 years, but in the 584-day Venus cycle, it would calculate out to be 2,340 years times the 584-day Venus Year. The 52-year period that the Mayan astronomy priest monitored for the Sun and the Moon cycle to coincide would fit into our 11,700-year half Precession 225 times. This is interesting because it also takes 225 days for Venus to revolve around the Sun. The rotation of Venus is 243 days, the number 225 of the Venus cycle, and the numbers of steps in the pyramid of the Sun at Teotihuacan (225 steps) may suggest a cosmic relationship between the Sun and then Venus cycles.

These rhythms that we are counting are like a spider web with different sized spaces on the web, but the whole spider web is resonating together and holding as one unit. We present a lot of numbers, but some of these are for more advanced studies. We do not expect anybody to memorize these numbers or to rehearse them. They are just here to show what we would begin to be counting, if we were the young astronomy priests.

Since the Mayans were counting the sacred 52 years, when the

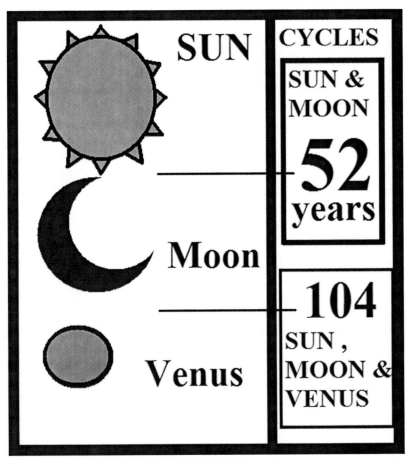

Figure 11-1: The brightest calendar measures in the sky, Earth, Sun, and the Moon align every 52-year cycle, and Venus will join every about 104 years in happy, family reunion as part of the fellowship of the Magical World Tree (Drawing by Gaspar & Jaramillo)

Sun and the Moon meet in one point, and they equated that to an alignment point in the Pleiades (January 1st on the Dendera Zodiac) and Venus, thus we understood that the years of 52 (Sun & Moon) and the 104 year cycles (Sun, Moon &Venus) needs to be part of the Cosmic Calendars. In the Scythian legend of the Golden Elk, that this is what the Magyars were counting. That 52 brave hunters went to chase the Golden Elk means to us that the 52 years for the Sun and the Moon that met in Orion, above the Belt that is called the Deer Head constellation by the Hindus, we find the recording down of important markers from the Mayans and the Magyars to place the meadow of the initial destruction into the great Hunter star constellation.

The arrival of the Venus to this important meeting point of the three brightest celestial bodies celebrated the 104 years. Mythologically, the two sons of Orion, the Hunter with their 50 soldiers met Two Princesses (Venus Transits) and the girls also had fifty maidens with them to add the total number of human participants to 104, matching the important cosmic number. That is 104 people in the legend of the Golden Elk. Subtracting the Four Royal members from the 104 would leave us with One Hundred, that is the number of Leagues hiding in between the Two Pillars of the Earth.

Thus, it is no wonder that the planet Venus was the main obsession for the counting of the long calendars. We learn about the Venus counts in the book by Gilbert and Cotterell, titled *The Mayan Prophecy*:

The closest astronomical calibrator is Venus, whose sidereal rotational period amounts to 584 days. 117 passes of Venus in the sky (117 x 584) amount to 68,328 days.

According to Gilbert and Cotterell, this is near the number of days that is 68,302 days, which amount to one complete sunspot cycle. When we multiply this number of 117 by a factor of a 100, it will yield a number 11,700 years, which is the length of a half Precession of the Equinoxes. Here we need to emphasize that several different cycles are being counted by the different calendars of the Mayans. The actual length of a solar year is about 365.25 days. The Mayan solar year counted a 360-day cycle, and also one of the wheels of the important T'zolkin calendar was established on the basis of a 260 days cycle. We do not want anybody to equate these or to be confused about the 584-day Venus cycle that was calculated separately from those cycles.

When we multiplied the sacred number of 2,340 x 5, we obtained 11,700 years, which is in the 72 years period of building the celestial temple in the active danger zone of the half cycle Precession of the Equinoxes, according to our theory. Thus, two of these half cycles add up to 23,400 years, marking the Full Cycle of Precession of the Equinox. It is called the 23,000 year Ice Volume Collapse Cycle in science. This is the actual length of the Precession found in the Ice Age drill-down in Antarctica. The Precession is not the wrongly calculated and insanely popularized 26,000 years of the Great Year of Plato. That is a very annoying and persistent mistake being repeated by famous scholars. By keeping a faulty count of the total, how can one add up the parts to equal?

For proof of advanced Mayan astronomy, we turn to the known numbers we have discovered and to the ones the Mayans advocated in their calculations. Thus, multiplying the 2,340 Venus number with the 584 day the synodic cycle of Venus equals 1,366,560 days, which is a well marked special number that refers to Venus in the Dresden Codex. We definitely think that the five cycles of 2,340 years of Venus providing the period of 11,700 years mark an important station in the shift of the ages.

Another interesting discovery that we found was on **Lord Pakal's TOMB that has 16 signs** around the lid, and these signs are signs that we know are part of the 260-day cycle of the sacred calendar of the Mayans. **These 16 signs are: 8 Sun, 6 Flint, 4 Earth, 5 Earth, 7 Owl, 9 Deer, 7 Sun, 11 Serpent, 2 Road, 2 Death, 3 Monkey, 4 Dog, 2 Sun, 13 Serpent, 1 Sun and 13 Death**. When we add the kin numbers of each of the sacred signs, we found a very important code to when exactly the next shift of the ages can take place on Earth:

The Sixteen Sacred Signs around the lid of Pakal's tomb:

South Side of the Lid:

The World Tree

8 Sun kin 60

6 Flint kin 58

4 Earth kin 17

East Side of the Lid:

5 Earth kin 57

7 Owl kin 176

9 Hand kin 87

7 Sun kin 20

11 Serpent kin 245

2 Road kin 132

North Side of the Lid:

2 Death kin 106

West Side of the Lid:

3 Monkey kin 211
4 Dog kin 30

2 Sun kin 80

13 Serpent kin 65

1 Sun kin 40

13 Death kin 26

We noticed that when we add all the kin numbers of each of the 16 sacred signs of the T'ZOLKIN count, we have 1,410, and this number probably is telling us that, from the time of the death of Lord Pakal to the end of the cycle will be 1,410 years apart, so that number is taking us from the day of March 23, 603 to the time between February 21 and March 31, 2013 (2013 - 603 = 1,410). Furthermore, from the moment Kan Bahlum II closed the tomb of Lord Pakal until the moment of the end of the cycle, will be exactly 1,320 years. We noticed that resembled the sacred Mayan calendar numbers of 13:20 and by playing with it and reversing the 13 and 20 it will yield 20:13, the year of 2013 likely just a coincidental number play.

There is another exciting numerical connection we found with the number 2,340. This we obtained from the numbers 130, the numbers of beads in the necklace of Lord Pakal, and the 18, which is 9 + 9 signs on the west and the east side of the lid of Pakal. We now know that the movement of Venus goes from east to west; that's probably the message behind the 9 signs on the east and the 9 signs on the west of the lid, which are revealing to us an important Venus code. When we separated the number 2340, and we cubed each number, interesting mathematical and numerical facts were found in the equation:

Possible hidden messages behind the rhythms of sacred numbers:

2340: we add it first (2 + 3 + 4 + 0) = **9**

(2 x 2 x 2) = 8 (3 x 3 x 3) = 27 (4 x 4 x 4) = 64 (8 + 27 + 64) = 99 (9 + 9 = 18) = **9**

(9 x 9 x 9) = 729 (9 x 9 x 9) = 729 (729 + 729) = 1458 (1 + 4 + 5 + 8 = 18) = **9**

(1 x 1 x 1) = 1 (4 x 4 x 4) = 64 (5 x 5 x 5) 125 (8 x 8 x 8) = 512 (1 + 64 + 125 + 512) = 702 (7 + 0 + 2) = **9**

(7 x 7 x 7) = 343 (2 x 2 x 2) = 8 (343 + 8) = 351 (3 + 5 + 1) = **9**

(3 x 3 x 3) = 27 (5 x 5 x 5) = 125 (1 x 1 x 1) = 1 (27 + 125 + 1) = 153 (1 + 5 + 3) = **9**

(1 x 1 x 1) = 1 (5 x 5 x 5) =125 (3 x 3 x 3) = 27 (1 + 125 + 27) = **153** (1 + 5 + 3) = **9**

So, when a number repeats itself, it is basically the result of cubing. The number in this case was 2,340, and the final result is 153. Now, let us take a look of this interesting game of numbers:

All the numbers that we have found when adding were 9, and this number repeated 7 times, so if we add the result of 153 to (9 x 7) = 63 we get the number (153 + 63) = 216, which again gets 9 (2 + 1 + 6), but the interesting thing is that when we multiplied 216 x 2,340 = 505,440, the number we get is basically the remaining numbers of days to the end of the cycle. This is because the number that we found in the Dresden Codex of 1,366,560 days and within the lid of Lord Pakal, and we add this number of 505,440, we get the number 1,872,000 days; that is the completion of the Great Mayan cycle of 5,200 TUN years or the 5,125 solar years.

Another interesting connection is that when we add the 16 signs by their position in the sacred calendar, they all add to kin 210 which is the energy of 6 Dog:

Sun = 20	Human = 12
Flint = 18	Death = 6
Earth = 17	Monkey = 11
Earth = 17 Owl = 16	Dog = 10 Serpent = 5
Hand = 7	Sun = 20
Sun = 20	Death = 6
Serpent = 5	Sun = 20

The two signs in the north-east side of the lid are 11 Serpent and 2 Road; if we added together the tones or energies (11 + 2) = **13**, and the sign on the north side, 2 Death, so that when we multiplied these two numbers (13 x 2) = **26**, we notice that on the north-west edge are

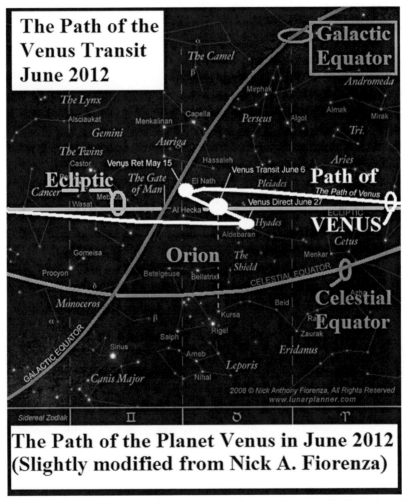

Figure 11-2: Venus will align with star RIGEL on June 4, 2012, and will align above with Orion on August 6, 2012, the very first day of the five unlucky days when the mythological gods were born in Egypt. (Picture by Fiorenza & Jaramillo)

The World Tree 261

the signs 3 Monkey and 4 Dog. When we add those tones or energies (3 + 4) = **7**, so very interesting is the fact that these two numbers match with the date of the last shift of the ages, recorded in Egypt on July 26 of the year 9792 B.C. It is not a coincidence that one of the streets in GIZA that takes you to the pyramids is called "July 26." From this mathematical code, we start thinking that the next shift could occur when the Sun will be in CANCER at the end of July or at the beginning of August, probably by the time the Mayans celebrate the five unlucky days of the HAAB calendar, August 6–10. We do not know, but we start understanding the meaning of all the cosmic players.

Another interesting fact on the lid of Lord Pakal's tomb is the number 144. This was well described by Mr. Cotterell. The number 144 was marked on Lord Pakal's forehead of. This is another piece of the puzzle of how accurate the traditional Mayan calendar could be. Alberto Ruz discovered the tomb on June 15, 1952, which in the Sacred Count of days was 13 Owl Kin 156. This day is the last day of the 13-day count of the Seed Kin 144. So, really, we can understand how powerful the Mayan calendar is as a tool to understand all the prophecies of what is going to happen at the end of the cycle.

Let us quote again from Mr. Cotterell's book to understand these numbers:

> *Look again at the decoded picture of Lord Pakal from the lid of Palenque. He carries the mark of the living god 144,000 sealed into his forehead. (Do not be surprised by the use of modern numerals, the super-gods could see the past, the present and the future)."*

We know that the Maya used fractal numbers to describe long periods of divine time, so as in the Bible one year can equal a thousand, thus the fractal number of 144,000 is 144, which is the kin number of the day sign of the Seed. This Seed day sign is in the exact thirteenth galactic week, when Alberto Ruz discovered the tomb of Lord Pakal, in the very last day of the 13-day count, on the day 13 Owl kin 156.

We learned from *The Orion Prophecy*, by Patrick Geryl, that the last shift happened on July 26, 9792 B.C. This number is within the lid by the multiplication of the sign in the north side of the lid, which is 2 Death and the two day signs in the east, which is 2 Road and 11 Serpent (2 + 11 = 13); this is 26 (2 x 13), and the number 7, which comes from the west side of the lid from the day signs 3 Monkey and 4 Dog.

Also, we have within the lid the date of February 21 and this date can be also interesting to watch in the near future in the years of 2013 or 2014. It falls near the date of the Gemini Twins, the celebration of the Valentine's Day and another important event around mid February when a more exact alignment is postulated with Cygnus.

Pakal has some of these dates in the north side, with the day sign of 2 Death, which stands for February, and in the south side of the lid, with the signs of 8 Sun and 6 Flint or Mirror. The two days signs in the west side of the lid are 3 Monkey and 4 Dog. Adding all these four signs equals 21. Do these numbers carved in stone carry ancient prophecies of how our living cosmos operates? Do they point to something else? Can they be calculated with a different mindset at hand? Numbers are helpful tools that can become sometimes double edged sword that they may yield numbers, sums and derivatives that may have not been intended. That is why, we operate within the realms of natural cyclical numbers found in the living Nature.

According to Patrick Geryl's theory, there was a shift on July 26, 9792 B.C. The about 11,800 year period approximates our calculations for the half cycle Precession that happened last than 12,000 years ago. Could another shift of the ages happen in a similar manner in the next few years between 2012 and 2014?

There is a 2013 February date in the traditional Mayan calendar that will be the Year 1 Wind kin 222 of the day 1 Wind kin 222, exactly 260 days after the Venus Transit of June 6, 2012. Now we can begin to feel the importance of the Venus calendar of the Mayans. The planet Venus was a special celestial object to watch in the heavens. That is why the Mayans dedicated five full pages of the Dresden Codex to this particular planet. According to the ancient star priests, at certain intervals something exciting happened when Venus occupied a special position in the sky.

The importance of the sacred Venus number of 2,340-year cycle is related to the special number of the Dresden Codex of 1,366,560 days (2,340 x 584 = 1,366,560). It behooves us, to better understand the

The World Tree

special positions of the planet Venus with its 2,340-year cycle, along with the famous 5,125-year Mayan Long Count cycle of the Earth. The 2,340-year cycle is very revealing to us, because it takes 5 cycles of 2,340 years = 11,700 years; that is our prominent Half Precession of the Equinoxes (11,664 – 11,736) when the Earth goes into a major cosmic transformation. This elevates the planet Venus to be a special player in the shift of the ages as a dominant counting device.

Definitely, we are living in very amazing times on this beautiful Mother Earth. Most of us feel that we are very close to some kind of shift in nature and consciousness happening on this Planet Earth. This may be something so exciting that it is beyond our human comprehension. Needless to say, that we sense it, and we begin to discover that all ancient cultures around the Earth were connected by these calendar

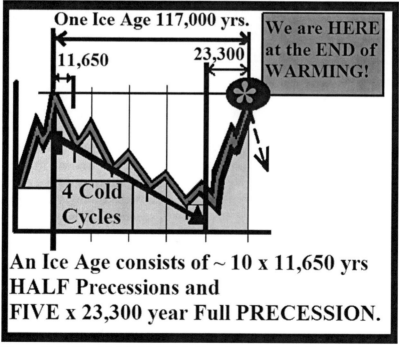

Figure 11-3: The 11,650-11,700 half cycle and the ~ 23,300 full cycle Precession that was popularized as the rounded mythological figure of the 24,000 years (Drawing by Wm. Gaspar)

cycles of nature.

The natural numbers of about 23,000 year long recurrent Ice Volume Collapse cycles were popularized by the world-famous scientist Wallace Broecker, who brought us the Atlantic Conveyor Belt Circulation Theory. These are the cycles that were predicted in the 1920s by the excellent Serbian mathematician, Milutin Milankovitch and these are the numbers taught in the Universities.

The Code for the Shift:

We discovered that in the Egyptian *Book of the Afterlife* and in the Bible, there is a sacred mystery behind the number EIGHT, which leads up to major disasters. Usually in the Bible, it is the Flood of Noah. In the *Book of the Afterlife*, it is the Judgment Hall of OSIRIS that is the gateway to the Netherworld. One might remember that the biblical Adam takes exactly eight generations to get to the time of Noah, and there are eight people on the Ark.

We can also detect the number eight as Plutarch describes the old romance of Cleopatra of Egypt to the Cesar of the Greco-Roman Empire. This was certainly an attempt to tie the Egyptian rulers to the Roman Emperors and to pass the mythological responsibilities of keeping the sacred stories alive from the ancient Egyptians to the related Roman Latinos, who at that time were developing the greatest and most influential civilization the world had ever known. In these events, we observe Julius Cesar and Octavius, who later declares that he is the Son of the Divine and changes his name to Augustus Cesar. This name will become one of the twelve months of the year. Also, the month of August currently the "eighth" month of the year, the month of Leo! The accounts of history takes on parts of the Solar Mythology as both of these handsome and powerful Roman rulers fall in love with the Egyptian Queen Cleopatra, whose early demise comes by the bite of an evil snake. Is it really the romantic history tainted and knowingly infused with the secrets of the universal cosmic mythology?

Interestingly, after this powerful and likely symbolic relationship of the Roman Emperor to the Egyptian Queen over 2,000 years ago, we can observe the emergence of the collection of books today we call

The World Tree 265

the Bible. It is not publicly obvious, but it is a very likely scholarly certainty that the ancient myths from the Egyptians, Greco-Romans, Hebrews, Hindus and Babylonians were cleverly collected into one holy book. Thus, the Bible is mainly a translated astronomical Egyptian Mystery School teaching brought to us by the combined wisdom of the Hebrew rebels and the bulging Roman Empire and then merged with the existing teachings of the pre-Christian Mithras.

Thus, the Sacred Cosmic Marriage of the Babylonians and the Egyptians had to be interwoven with the Bull killing rituals of the Mediterranean cultures and tie it to the Twelve Labors of Hercules. Not an easy task by any means to satisfy the varied needs of the multiethnic Roman Empire. A worthy parallel would be to try to merge the multiple religions and spiritualities practiced in the United States today. It would have to be an unattainable feat just to have the Catholics and the rest of the Judeo-Christians to agree on the specifics, then to try to mix into this the Buddhist, Hindus and the growing number of Muslims is unimaginable. Above all, if one would have to satisfy the spiritual needs of the original aboriginal populations of the Native Americans and the Nature worshippers then the requirements would be extra ordinary. We should say that it would not be impossible, but it would have to involve an Act of God.

As matter of fact, that is what the earth changes will achieve in a few decades if the Mayans are correct. The unimaginable power of our Creator that can move the mountains out of its places and shake the globe as a table cloth, shall convinces most of us that there is only One Creator Force in this Galaxy that is strong enough to move rocks by the means of the cosmic wind. Intense theological arguments about who fits into the Holy Trinity or whether eating fish on Fridays are allowed or not will not be needed, because the One who sends the intense Fire, the harsh Water and the intense Earth quakes will not care how we call Him or Her or whether we eat or not, as those are not questions in the Spirit realms of existence. For our Creator, the abundance of the entire Universe is so vast and everlasting that we doubt that there is any punishment involved in these charging up the Earth's geo-dynamo.

Now, let us return back to the number eight. From the Egyptian teachings, we need to identify those crucial eight gods, or the generations leading up to the punitive judgment of Nature. Erik Hornung shows in his book, *The Ancient Egyptian Books of the Afterlife*, that within that collection of writings, there is an enigmatic Book of the Afterlife, a Book of the Amduat, and also a Book of Gates. These works contain symbols that are not translated into the regular accepted picture

Figure 11-4: The Egyptian God AMUN-RÁ who represents the Unified RÁ / RÉ (Sun King) and OSIRIS (Orion) is standing between the Two Sisters; Isis and Nephthys that represent the Earth Axis, THE TWO PILLARS and connected to The Three Sisters concept of the Venus Transits (Location: Tomb of Nefertari, Valley of the Queens, Thebes, Egypt. Photo Credit: © DeA Picture Library / Art Resources, NY)

hieroglyphics. In Hornung's book, we can observe the Eight Gods lining up to be judged. There are two goddesses standing behind the mummified pharaoh, who sits on his throne. These two female deities are the Two Pillars of the Earth and in our secret formula, closely tied to the Venus Transits of the 96 pairs, and the animal gods lining up to be judged are the astronomical time lines of those years when the shift happened.

The Ram-headed AMUN-RÁ, the in itself represents the westerly Uniting Galactic Spiral Forces by his Horns, is shown between the Two Pillars of the Earth. Remembering back to the first two chapters, we learnt about the SA-RA concept connecting the west (SA – Goose) and the east (RÁ-Sun) forces, we can learn another way to relate to that same exact cosmic science. Just as we study the body of the Egyptian Lamb of God, Amun-Rá we can notice that the Horns stand for the Heart of the Galaxy in the west where the Two Spirals unite, then he has the Sun positioned on top of that unified horns and that certainly means the easterly Sun. Therefore, since we know that this Egyptian Lamb of God is the Three Months period from December until March, we also recognize that the combined Hero we are dealing with is ORION. Now, when the unification begins in the star constellation Orion we can also go up a notch and find the other Club-yielding Hero Hercules ready to close the Hour between the Magnetic North and the Polar North, thus the Hercules Hero from the North becomes united with the southerly Hero of Orion.

When we reach this Fourth and Fifth Hour of the Last Twelve Hours we reach the westerly ZERO TIME between the two spirals of the Galactic Center. The Zero Count of the Last Twelve Hours in the case of the Earth count becomes synonymous with the Zero Count of the West Four Years later. Maybe this is another reason why the name of the Resonating Northern Cross in the Heart of the Galactic Center is named 'Négy' meaning 'Four' in ancient Magyar.

The head dresses of the Two Ladies (other times represented by the Serpent and the Vulture combination on the head of the Pharaoh) are again very revealing. The Magnetic North Lady standing on the left of Amun-Rá wears a slab on her head that contains the Red Brick of the Core of the Earth. On top of the slab is usually a green 'Kup' that immortalizes Mother Earth cut in half on the Galactic Plane. This is also the originator concept of the Holy Grail. When the Magnetic North tips toward the south, the Blood feels up the Kup. That same Kup is the equivalent of the letter 'K' in the Egyptian Hieroglyphic Alphabet. Now it is even clearer, we hope why is it important to use the letter 'K' instead of the letter 'c' when it is cosmically appropriate.

Further, the Belt on both ladies (not exactly the same way as on Amun-Rá!) is shown as an upside down pyramid that is met at the Knot with a phallic symbol fashioned out of the two hanging ends of the belt. They say, sex sells, but in this case it is modeled in a subtle fashion. Needless to say, the colors of their outfits are made up of only Two Colors, the Red and the White! Well, we certainly understand that those Egyptologists and students who studied these pictures until now are in total shock and upheaval, realizing that until now that teachings that they received have not mounted up to the teachings of the Rá Priests.

The Lady on the right of Amun-Rá stands for the Polar North and the Throne of Isis (Cassiopeia) already gave away her astronomical position. The Red Ribbon tying up their hairs have still Two Endings, meaning that although the Cosmic Sex has begun and the cosmic sex organs are near to each other, the Axis shift has not been completed. Compare this to Queen Nefertari, who prays alone on the left side of the picture and has both of her arms lifted parallel, that means the Two Pillars of the Earth had become unified and the Axis shift has been completed. Her parallel Two Large Ostrich Feathers say the same and the Vulture in place of the split Ribbon foretells death. Exact Egyptian Cosmology is not taught at any of the universities, because they are not aware of any of these explanations.

Furthermore, we have another unsolved cosmology to share. On the left middle side of the picture one can notice a spotted cow. The spotted aspect of any human or animal represents the 'Boils of Egypt' mentioned in the Bible and it is commonly mistaken for Leprosy. The spotted animals that Jacob owns in the Old Testament are certainly not infested with leprosy. The spots mean a much more sinister medical scenario. Here we again call on the knowledge what we learnt during the studies about the Mayan Brothers. The dark spots on the Mayan hero actually mean 'Sun Burn Marks'. That translates into such a strong force of the piercing Sun that not only it will create boiling rivers, but it will burn boils into the skin of humans.

The boil on the skin and the boiling water in English provides an interesting linguistic clue. In this same cosmic scenario, the bards from different continents maintained a very rigorous symbolic discipline to pick animals that are naturally 'spotted' around that time of the sky, which are the Camel, the Lynx Cat, the Serpent and the Leopard or the Jaguars.

None of these interpretations have ever been heard before in the university halls of any institutions and we certainly expect a backlash of the academia who so long benefited from teaching the misunderstood science of Egyptology or Theology. We also thoroughly understand that not many priest, shaman or medicine man would want to bring the 'bad news' as we are presenting here. Although, these interpretations of ours could be very disturbing, we feel that it is not any more frightening than what we can see in the Holy books of Judeo-Christianity and Islam. We do not blame an angry god, we only

Figure 11-5: The young pharaoh sits on the throne facing the Hawk, this time. He sits between Two Sisters, who represent the Two Pillars of the Earth and connected to the Venus Transits in front of the Sun, which happen 8 years apart (Drawing by Wm. Gaspar)}

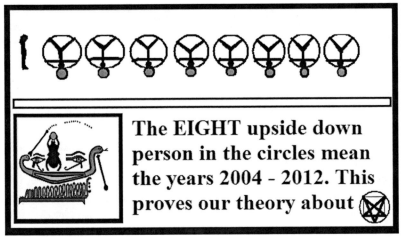

Figure 11-6: The PORTAL that is the last VENUS TRANSIT PAIR of the 96 pairs followed by the 'Face of God' that is the Great FLOOD in the Tomb of Ramsses IX. (Modified from Erik Hornung's book titled The Ancient Egyptian Books of the Afterlife by Wm. Gaspar)

try to understand why Nature has to cause total destruction to allow rebirth. Maybe, the flesh existence is truly a lower level of existence in a density bound by gravity and a physical tragedy only opens the gates of heavens for us to return back to the all knowing Spirit realms.

While we are in this dense existence, we would still rather bring good news, but if it is bad news we would like to know exactly what to expect potentially in the near future. It might not happen, but if it does and one would like to physically prepare for it, then we might as well consider the worst case scenario as the blue print for the meaningful flesh preparations.

The first three god figures on the upper level Judgment Scene are the Goat (CAPELLA), the Cat (LYNX), and the Lion (LEO). These would coincide with the astronomical north alignments to Polaris of the current dates from January 26, the Chinese New Year almost until the time of Easter (The Lamb is shown as third on the bottom row). Thus, that seems to justify our theory that the first few large eruption of the Sun at the end of the Fourth Hour and the beginning of the Fifth Hour will graduate by the alignment of the Goat Capella star in the Chariot star

The World Tree

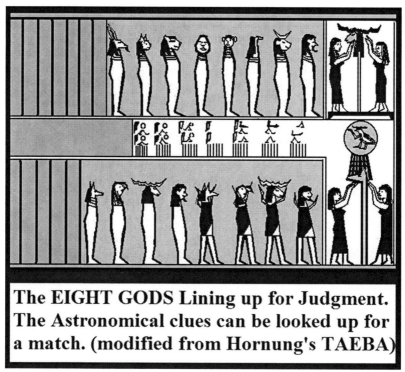

Figure 11-7: The Gods before Judgment (Modified from Erik Hornung's, The Ancient Egyptian Books of the Afterlife. Drawing by Wm. Gaspar)

constellation. The next station in this hellish scenario is the Lynx Cat who is assigned to cut the head of Hydra. Then the third astronomical animal, the Lion will complete the first stage of the axis shift and will rule Mother Earth for a few years perhaps. The forty years of exile – again understood in the realm of cosmology – suggests that whatever happens to the wandering path of the axis of the Earth, it will not settle semi permanently until the forty year period ends. Adding this long forty year stretch to the 32nd year of the 36 year path, will provide us with the 72 years to build the temple. These seventy two years is the One Degree Precession and it is more than understandable that the priests tried to create a sum of numbers that could be and should be remembered.

The eight gods on the upper level are shown from left to right, beginning with the Goat, then progressing with the Lynx cat, the Lion,

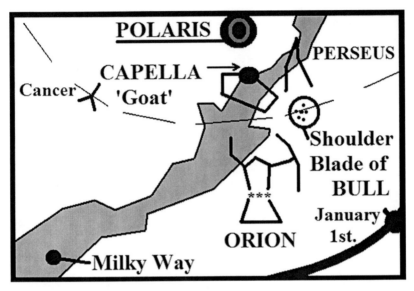

Figure 11-8: Orion is in the center at the end of January, when the infamous CAPELLA Goat star in the Chariot of the Fire / Sun star constellation lines up in the sky (Modified from the Star Talker Astronomy Charts.)

and the Bald head / face turned 90 degrees. Next is a human head with hair, followed by Snake / Vulture. To the right are shown the Bull / Cow and the Male God. If one counts on amongst the captives then the numbers will add up to the forty years. What forty years of displaced axis will do to the health and prosperity of the Earth and the survivability of humans under those conditions is difficult to imagine.

Religious and mythological tales identify the time between Christmas and Easter as when the ancient earth changes happen. The Star Talker Charts are aligned with Polaris, the Northern Star, in the center, just as the Zodiac of DENDERA of Egypt is showing the alignments. These animals are not random animalistic pictures as so many scholars around the world assume; they are very specific messages of what happened to the Earth. The horns of the Goat have different meanings from the horns of the Ram. The Goat with the horns apart shows the two North axes apart. The Ram is more than that.

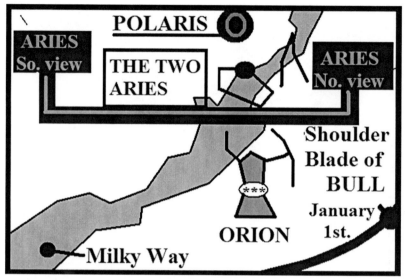

Figure 11-9: The Two RAM. The Aries sign in this alignment ends around Christmas and begins at Spring Equinox. The winter months between the two Aries Rams hide a telling mythological tale of an ancient disaster. One spiral horn switches to the other! (Drawing by Wm. Gaspar)

The Donkey is not only the animal symbol of the American Democratic Party, but is also grouped together in the book of Job with several other creatures. In the next quote, the Goat is mentioned. It is the first of the eight, and the Donkey is part of the Crab, below the astronomical Lynx in the sky, as we learn later:

Do you know the time when the **wild mountain goats bear young**? Or can **mark when the deer give birth?** . . . **Who set the wild donkey free**? Who loosed the bonds of the **ONAGER [Asian wild Ass]**? . . . Will the **wild ox** be willing to serve you? . . . The wings of the **ostrich** wave proudly, but are her wings and pinions like the kindly **stork**? . . . Have you given the **horse** strength? Have you clothed his neck with thunder? Can you frighten him like a locust? . . . **Does the HAWK fly by your wisdom, and spread its wings toward the SOUTH!"**

—Bible Job 39:1-16

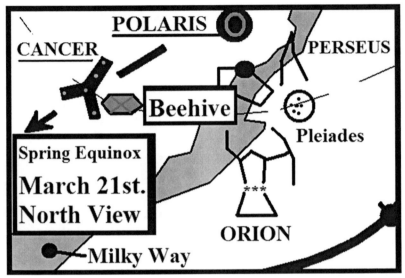

Figure 11-10: Around the Spring Equinox of March 21 is the alignment of the Beehive and the Cancer star constellation that contains the Two Donkey stars of the Earth Axis shift at the Birth of the Son of the Sun (Modified from the Star Talker Chart by Wm. Gaspar

The GRAIL star constellation aligns around the time of Easter. Cancer is before that, at the Spring Equinox, around March 21, or mid-July, if the southern look at the sky is considered. That is the approximate time when the Lynx Cat using the pincers of the Cancer Crab cuts off the head of Hydra.

The wide variety of these animals herded into only one chapter of the book of Job suggests to us that Job knew more about the symbolic nature of the creatures than one could easily decipher. We did not spend much time discussing the horse earlier, but it is enough to say that it parallels the actions of the Two Rams from Pegasus to the Unicorn and the Centurion. Sagittarius, which stands for the mathematical westerly Center of our Milky Way Galaxy. By current scientific estimates, it takes 200 million years for our galaxy to take a full turn. The ancients were well aware of that. Their knowledge is reflected in the statements about the "two hundred million horsemen" who are ready to do battle in the

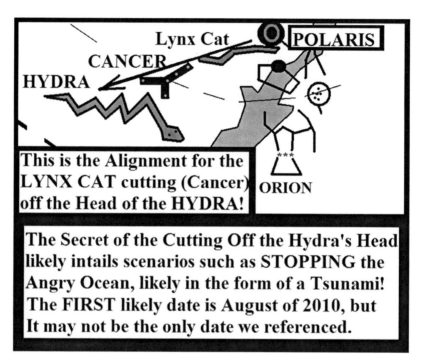

Figure 11-11: *The Spring Equinox is the time of the alignment of the Cancer constellation and the Beehive star cluster in the northern system or the Zodiac of Dendera (Drawing by Wm. Gaspar)*

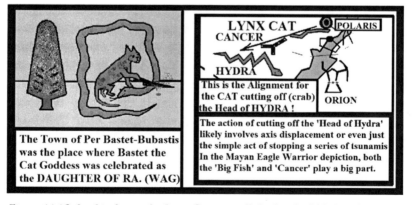

Figure 11-12: *In this figure, the Lynx Cat cuts off the head of Hydra, the Water Snake, using the two edged sword of the Crab (Drawing by Wm. Gaspar)*

Bible. This, we mention as another proof for the advanced scientific knowledge base of the so called Atlanteans.

Thus, by the wisdom of the Ancients we are suspecting that if the first warning of a bright light happens on the night of December 24 – 25th of either 2012 or 2013, then we have about four weeks to prepare for the fiery insult. Thus, our first animal of the last eight ties the Goat to the Chariot of Fire, and to that fateful, tragic year in the past that may repeat itself in a few years. The goat star, Capella, is not only very important for the conclusion of our cosmic story, but it was essential for the survival of the baby Greek god, Zeus. The 'She-goat' was called Amaltheia by the Greeks and earned her fame as the wet nurse of Zeus while he was hidden as a baby in a cave on Mount Ida.

Capella in Latin refers to Little Nanny for her role of feeding the baby god. Later, Zeus used her hide to make the famous thunder-shield (Aigis), which is naturally important in our astronomy, since the Shield of Orion lines up with the Goat star! The horns of Capella, then, provided the cornucopia, the horn of plenty (Keras Amaltheias). In "The Sibylline Oracles" of the Jews, from Barnstone's *The Other Bible*, the following quote is on page 504:

Then mortals in desperation, in the last stages of famine, will devour their own parents will consume them greedily as food . . . As a result of grievous wars the bloodstained ocean will be filled with flesh and blood of insensate men . . . I saw the threatening of the gleaming sun among the stars and the moon's grievous wrath among the lightning flashed. The stars travailed with war and God suffered them to fight. In place of the sun, long flames rose in revolt, and the two-horned revolution of the moon was changed. Mounted on Leo's back Lucifer waged battle. Capricorn smote the heel of the young Taurus and Taurus snatched the day of return from Capricorn. Orion removed the scales so they disappeared. Virgo changed her sphere with the Twins in Aries. The Pleiades no longer appeared and the Dragon disowned the belt. Pisces entered into Leo's girdle. Cancer did not stay for he feared Orion. Scorpio drew up his tail, because of savage Leo, and the Dog Star perished from the Sun's flame."

In the above quote we receive a lot of astronomical information about the Earth changes, but it is only with our Cosmic Model and the Axis shift theory that one can make sense of it. Orion, the Sun's flame, Leo and Capricorn are all explained in our theory.

Even the Hindu Myths tie the fire destruction to the goat:

Then the triple world was totally destroyed by the fire of SHIVA'S anger . . . And even when everything had been destroyed, the noble lord who is an ocean of pity granted them safety, and gave the head of a goat to the man DAKSA and revived him.
—Hindu Myths, page 251

Then the mightiest of the mighty saw his father AGNI approaching, and he honored AGNI, who remained there together with the group of Mothers, who was born of anger; she held her trident in her hand and protected him as if he were her own son. The cruel daughter of the ocean blood, the drinker of blood, embraced the great general and cared for him as if he were her own son. And AGNI transformed himself into a **goat-headed merchant with many children**.
—Hindu Myths, page 114

As we learned earlier, SHIVA is the Sun and AGNI is the Hindu Fire God, and both of those are tied to the head of the goat. We noted the goat in the Egyptians, Jewish, and the Hindu myths, so now we can take a look to see if the goat or the other animals appear in the Bible and are tied to the number eight and to fire. This is how the Bible writes about the animals:

And the Lord spoke to Moses *saying:* **When a bull or sheep or a goat is born,** *it shall be seven days with its mother, and from the* **eighth day** *and thereafter* **it shall be accepted as an offering made by fire to the Lord.**
—Bible, Leviticus 22:26-27

Not only the "Goat is born" in the above quote from the Bible, but it is tied to the number eight and to the offering made by fire to the

Lord. Is it only a coincidence, or is our theory correct, and even the Bible contains mainly astronomical clues?

> *Then he shall take a censer full of burning coals of fire from the altar before the Lord . . . He shall take some of the **blood of the bull** and sprinkle it with his finger on the mercy seat on the east side; and before the mercy seat he shall sprinkle some of the blood with his finger seven times. Then **he shall kill the goat** of the sin offering.*
> —Bible, Leviticus 16:12

All of this is tied to a burning fire at the altar of the Lord. It becomes clear that the Semitic science priests of Ra brought us the Egyptian Mystery School teachings through the Old Testament of the Bible.

Certainly, the degraded morals of the declining Greco-Roman Empire, along with the mixing together of several ethnic groups with differing moral values within the immense borders, necessitated the combined emergence of religious law that preached morality, but hid science and astronomy. This progressively more moral, but less educated hard working agricultural population pried the proverbial doors wide open for the Dark Ages to march into Europe. Not until the emergence of science and enlightenment, about three hundred years ago, did the darkness of ignorance slowly started to dissipate.

Returning to Figure 11-6, it is important to realize that the top row figures are mummified and have no arms, which is to show that the king lost the right arm that is the 90 degree angle solar stabilization of the earth. The lower register shows the same four gods: the jackal (bear), hawk, ram, and male god for the first four years mummified, then the next four years with small upper extremities that means, to us, lesser electro-magnetic power to the poles.

The first four are wearing white outfits and the next four wear mainly dark, which may be alluding to the fact that, during the eight years of Venus cycle, we have the planet four years as a Morning Star and four years as an Evening Star.

A few more quotes will bring the fire, harlot, jackal and even the bald man (the fourth god in the row in Figure 11-6) from the Bible to our Egyptian round table of discussion:

> *All her carve images shall be beaten to pieces, and all her pay as a **harlot shall be burned with the fire**; . . . Therefore I will wail and howl, I will go stripped and naked. I will make a wailing like the **jackals** and mourning like the ostriches . . . **Make yourself bald and cut off your hair . .***
>
> —Bible, Micah 1:7-16

No need to be repetitious about the combined wisdom of the harlot, fire, owl, jackal, and ostrich and making oneself bald. The nakedness is tied to the Sunspot eruption because the intense heat made burns and boils on people's bodies and they were not able to stand the clothing on themselves. Therefore, "revealing or uncovering one's nakedness" in the Bible is not the action of a pervert human, but it is about the intense heat of this ancient cosmic event that happened 11,700 years ago.

Reviewing the ancient classical Greco-Roman mythology, we shall outline the labors of the famous HERCULES to see if there is a close correlation to the astronomical animal symbols of the Egyptians and biblical wisdom.

The 12 Labors of Hercules

Before the twelve labors, our hero, Hercules, became wild and, in a fit of madness, killed his wife and threw his three children into the FIRE, similar to the Goat or Chariot of Fire. After the fiery disaster, he was ready to start the famous TWELVE LABORS:

1. In his first labor, Hercules slew the NEMEAN LION. (Both the month of August and the Age of Leo).

2. During the second labor, our hero killed the seven-headed serpent, the HYDRA of LERNA. During this act, he crushed CANCER, the CRAB. (Donkey and Beehive also!).

3. In the third labor, he captures the Golden STAG of Cerynea whose fleetness was such that he barely touched the ground. Only a very cold winter's snowfall helped in the capture the Stag up in the

North. (The astronomical "three skips of the antelope in the sky" and the severe cold winter that followed the sunspot eruption).

4. Overcoming the wild BOAR of Erymanthus in Arcadia happened during the fourth labor. During this task, Hercules mortally wounded his beloved tutor, CHIRON, who became the constellation SAGITTARIUS.

5. In his fifth labor, Hercules was appointed to clean the filthy stable of King Augeas of Elis. The giant had to divert the fast-running RIVER Alpheus to flush the barn then he guided the stream back to its old course.

6. The sixth labor brought the capture of the mad BULL of Crete that was given to King MINOS, ruler of the island, by Neptune, God of the Sea.

7. The seventh labor of Hercules involved feeding King DIOMEDES of Thrace to his own cannibalistic HORSES, who were then taken captive.

8. In the eighth labor, Hercules was sent to the land of the Amazons to retrieve a girdle worn by Queen Hippolyte. On the way home, the hero saved the Trojan Virgin, Hesione, from the jaws of the sea monster.

9. The ninth labor of Hercules saw the slaying of the metallic birds of Stymphalia that had come from the depths of the Earth, causing three years of DARKNESS over the stagnant water of the lake.

10. In this labor, he captured the CATTLE of GERYONES and slew a thief Giant named CACUS in a dark CAVE.

11. In the eleventh labor, Hercules had the task to find the Golden Apples of Hesperus, god of the west, also known as the Evening Star. A fierce DRAGON was guarding the apples. Hercules set Prometheus free also, who was still bound by chains.

12. In the last and most difficult labor. Hercules needed to carry the TRIPLE-HEADED DOG back into his cage to end the time of bondage. In this other pursuits, Hercules also helped Jason and the Argonauts and was part of the first siege of Troy.

Analyzing the labors of Hercules, we encourage all our readers

The World Tree

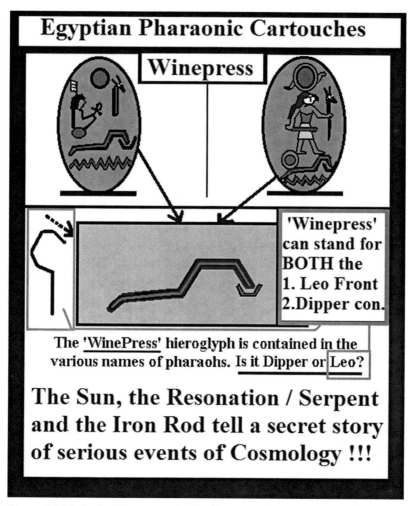

Figure 11-13: Is the Winepress the Big Dipper or Leo or Both? (Drawing by Wm. Gaspar)

to come up with their own discoveries and conclusions about the sequence of events of this ancient tragedy.

For now, let's go back to the importance of the number eight. The Venus Transits occur eight years apart, and in the middle of that period is when, according to our theory, the purifying Lord of the fiery evil earth changes appeared from the eruption Sun. Cleverly, the graphic nature of the number 8 represents a simple, never-ending eternity that includes two halves to the whole.

The last concept that we want to cover in this chapter is the dipper that contains water that can be poured out all over the Earth and the winepress symbol that the Egyptians used. It was an important discovery for us to find that both the Little and the Big Dipper on the astronomy chart, and the sickle-shaped Leo sign can stand for this elusive winepress. Thus, as obvious as it seems, shockingly the Little Dipper is not the final answer for the symbol of the winepress. How is this winepress related to our earth changes scenario?

Let us mention a few quotes from the Bible to understand the cosmic meaning behind the winepress:

A third angel followed them and said in a loud voice: If anyone worships the beast and his image and receives his mark on the forehead or on the hand, he too, will drink of the wine of God's fury, which has been poured full strength into the cup of his wrath. He will be tormented with burning sulfur in the presence of the holy angels and of the lamb.
—Bible, Revelation 14:9-10

Then I heard a voice from heaven say, write: Blessed are the dead who die in the Lord from now on. Yes, says the Spirit they will rest from their labor, for their deeds will follow them. I looked and there before me was a white cloud, and seated on the cloud was one "like the son of man" with a crown of gold on his head and a sharp sickle in his hand. The another angel came out of the temple and called in a loud voice to him who was sitting on the cloud, take your sickle and reap, because the time to reap has come, for the harvest of the earth is ripe. So he who was seated on the cloud swung his sickle over the earth, and the earth was harvested.
—Bible, Revelation 14:13-16

Another angel came out of the temple of heaven, and he too had a sharp sickle. Still another angel, who had charge of the fire,

The World Tree

came from the altar and called in a loud voice to him who had the sharp sickle, take your sharp sickle and gather the clusters of grapes from the earth's vine because its grapes are ripe. The angel swung his sickle on the earth, gathered its grapes winepress of God's wrath. They were trampled in the winepress outside the city and blood flowed out of the press, rising as high as the horse's bridles for a distance of 1,600 stadia.

—Bible, Revelation 14:17-20

These quotes are really important in our theory, because the first part of the book of Revelation 14:13-16 makes it very clear that the angel sitting on a cloud is the constellation Leo, and this constellation has the shape of a sickle, and it will be after the first shift to ripen the earth. It is clear that the second sickle is Ursa Major. This constellation rotates counterclockwise around POLARIS (which is the orientation of the World Tree or Axis Mundi) and is known as the Big Dipper. This will ripen the grapes of the earth, which are the blood of all human beings after the second shift and that's the one that will change the color of the water into blood by volcanic eruptions. That is what we think was the expression for the winepress in the Bible.

CHAPTER **12**

The Last Twelve Hours of the Hero's Journey

He who desires to attain to the understanding of the Grand Word and the possession of the Great Secret, ought carefully to read the Hermetic philosophers, and will undoubtedly attain initiation . . .
—Albert Pike, Morals and Dogma

Similar to other cultures, such as the Greco-Romans, the Hebrews, and a list of others, the Egyptians did not lack in festivities. We are confident to state that most of our current day holidays have their roots in some fashion amongst the Egyptian celebrations. There were occasions in Egypt to celebrate the harvest, the Union of the Two Lands, the Union with the Sun's Disk, the New Year, and the Divine Birth of the Goddess's Child. The celebration of the Sacred Cosmic Marriage was held at Edfu and lasted fourteen days. Two weeks before the new moon, the celebrations began in the town of Dendera, the headquarters of Goddess Hathor; on the third month of the season at new moon, the festivities began to shift to the Nile, and the Goddess was placed into a boat to be transported upstream toward the south. The lovemaking and drunken singing and music lasted from new moon until the full moon of that same month.

In a fashion probably not dissimilar to a Brazilian fiesta, the Greek Bacchus, god of wine celebrations, or the Mardi Gras of New Orleans,

The World Tree 285

Hathor, the beautiful goddess of fertility and sexuality was shipped to Edfu, to the hometown of Horus, where his temple was located. The parts of the wedding ceremony was reenacted by the ruling Pharaoh and his Queen. There in Edfu Egypt, the two, a god and a goddess, spent two weeks together in *merry* making. This Egyptian and Babylonian MERI/MARI word means "love or making love," that is, *Cosmic Sex*.

The Male God **'MIN'** was clearly represented on the walls of the pyramids with nakedness and an obvious phallic symbol. The god M-IN would be the 'IN-dividual' designation and the god A-M-UN then the 'UN-ited' aspect of this Egyptian Enigma. Therefore, the Egyptian male deity MIN became the masculine part of the MIN-ER-VA concept later in Rome. This Cosmic representation of the Male Phallus in Egypt was not done for the shock value of Vulgarity, but to represent the Male power of the Cosmic Union. It was representative of that single-day event when the Phallus of the Male god *symbolically erupted* or *prematurely raptured* and that was the day that changed the lives of billions of people almost 11,700 years ago. That one gigantic male eruption destroyed civilizations in a blink of an eye, and probably earned the male designation to the bi-gendered or genderless Creator Spirit.

Another related festival was held at the Temple of Luxor, or at Medinet Habu and it was designated the Feast of Opet that linguistically also relates to the Opera sign of the Venus Transit. **The Feast of Opening** was related to the **Opening of the Mouth of the Crocodile (Dragon)**, but by inference, it was also connected to the opening up the New Age by opening the Gate of Heaven, the Portal that was fashioned out of the last Venus Transit Pair. According to Lesko, there were food offerings of 11,341 loaves of bread and 385 jugs of beer. Together, it amounted to 11,726 pieces of food offerings, and is very interesting how Lord Pakal leaves these clues for us in his tomb lid. He was the eleventh king of Palenque, and in the north-west corner of the lid of his sarcophagus is the number 7, and in the north-east corner is the number 26. Also in the lid, we found 11 groups of 3 dots, so we can identify the sacred number of 11,726 in a very interesting Mayan-Egyptian connection. These numbers may represent the years between the time periods of major disasters on Earth.

The Book of the Dead illustrations based on the shape of the Pentagon to hide the secret of Cosmic Union. The Pharaoh wears white and the Queen is in red. Their belts are just the opposite. The

Figure 12-1: The Pharaoh and his Bride in a Pentagon shaped enclosure tied to the Sacred Cosmic Marriage of the Gods and the Judgment (Modified from The Egyptian Book of the Dead by Chronicle Books. Graph by Wm. Gaspar)

The World Tree   287

unification of the Red (Magnetic North) and the White (Polar North) is displayed in this Cosmic Marriage. She has the Vulture with a Cobra head for a Head dress and on top of it is the Throne. The Hawk on the roof wears White that means that the axis shift has happened and the Red Hawk is now wearing the White color of the Polar North.

Thus, if remember that most or all of these holidays and celebrations mask the sad earth change events. At least, for the next 12,000 years the people had a chance yearly to celebrate about something, that before in that particular year they were crying about. Most of our happy holidays are untold reminders of ugly earth change events.

The temple calendar of Medinet Habu are listed a few other feasts that were held there. These were the Feast of the Valley, the Feast of Amun, and the Feast of Lifting Up the Sky. This last feast reminds us of either the COSMIC BOAR or a return of the sky to its previous position after the initial displacement. Another major celebration, which lasted twenty days, was the Feast of Coronation. This was marked in the twenty-second year of the reign of the king. This celebration may be related to a celebration of the thirtieth year of the reign of the pharaoh Amun-Hotep. This gave an opportunity for the ruler to officially declare himself transformed into a cosmic deity, the Sun disk itself.

Interestingly, in the last eight years of his reign, when between 22 and 30 years of age, he is depicted much younger, as if he was reborn. That is likely the seven years of bounty followed by a pre-cataclysmic short period of 'teaching' as Jesus did in his last few years. By today's calendar, supposedly the 30 would fall somewhere between 2009 and 2010. It is difficult to decide when the 36 years the ancients keep showing commenced and when they ended. We assume that those 36 years span between 1981 and 2016, and in that period we also witness the beginning of the Last Twelve Years of the Hero's Journey, starting with the sacred year 2008 (zero year) when the planet Venus was in the dead center of the Sun.

The exact date was June 9th, 2008. The first year then would begin with December 21st 2008 (Winter Solstice), January 1st, January 26th or around March 21st, 2009, which would be the Spring Equinox. Examples can be found for all of these beginnings to a New Year and more.

Thus, if 2008 is the Year Zero of Venus and thereby the Year Zero of the Hero's Journey, then by simple calculations 2009 is the 1st year,

2010 is the second year, 2011 is the third year, and the FOURTH YEAR is 2012, all of which are shown in The Ancient Egyptian Books of the Afterlife by Erik Hornung as well watered years before the First Huge Sun eruption of the Fourth Year of the Last twelve in 2012.

Apparently, according to the Erik Hornung's book, between the Third and the Fourth Hour of the Last Twelve the severe Earth changes begin. This is how the book writes about it.

The presence of Osiris manifests itself a number of times in the lower register of the third hour, and in the text that concludes the hour, Re is even said to turn and face him directly. We also encounter avenging creatures with knives in hand to render all enemies harmless. This well watered, abundant landscape ends at the fourth hour. Here lies the desert of Rosetau, the "Land of Sokar, who is on his sand," a desolate, sandy realm teeming with snakes whose uncanny movement is emphasized by the legs and wings on their bodies. A zigzag route filled with fire and repeatedly blocked by doors leads through the region of this hour. For the first time, the solar barque needs to be towed for it to make progress, and the barque itself turns into a serpent whose fiery breath pierces a pathway through the otherwise impenetrable gloom.
    –Erik Hornung, *The Ancient Egyptian Books of the Afterlife*

Thus, we are certain that the Sun eruption comes in the Fourth Hour of the Last Twelve. The question, when is the First Hour? What incredible mile marker can we use to be sure that it is worthy for the beginning of those crucial Last Twelve Hours? Can we be sure that the Year 2008, when the planet Venus is in the dead center of the Sun, is the '0' year of the Last Twelve Years of the Hero's Journey? Well, if we assume that then the Fourth Year of the last twelve would fall on to the year 2012, the year the Mayans identified as the end of an old age and the rebirth of a New Cycle of the Sun.

We may have some warning even by the spring of 2012, although most likely on December 25th of 2012, on the Fourth Day of after the Winter Solstice of the Fourth Year of the Book of the Dead, when the Four Directional Northern Cross, called 'Four' ('négy') aligns with the

Sun - we shall have the first illumination appear in the sky. The fourth and most powerful Sun eruption will likely happen when we reach the Beehive cluster on the Ecliptic, although the eruptions in the past around January 23- 26th of the Fifth year evoked the images of the Devil with Goat head, as it happened when the Goat star Capella lined up in the mid-winter sky.

It is difficult to determine with any amount of accuracy what different cultures understood as the first year of those turbulent changing times. Also we need to remember that the Venus Transit of June 8, 2004, was an important marker in the heavens, thus we are very certain about that day as Zero Year or the First Year counting down to the Last Twelve. The Mayan or the Egyptian Astronomy Priests, being in total possession of the cycles of the planet Venus, would have not picked any other marker for that date, but the time when Venus was behind the Sun in the dead center. We know that because of the representations of the Bull with the Sun between the horns displaying an obvious DOT or POINT in the middle of the Sun. That was one of their openly displayed secrets. The Hero's Journey in the Babylonian Epic of Gilgamesh was also tied to the Bull. That Bull reference probably identified the '0' year of the Venus being in the center of the Sun, and also the Bull reference could be tied to the troubles that began in the Fourth Year when around January 1st the shoulder blade of the Bull, the Seven Sisters of Pleiades align on the Zodiac of Dendera.

The Sunspot eruption seemed to be after the third Venus Transit. Thus, four years after that central year of 2008, the Sun began to act up in December. Today that date would be December 25th, 2012 according to the Mayans, and it may even be a year earlier or later, such as in 2011 or in December 25th, 2013 according to the Egyptians. Why we are giving a two year span for the first Big Sun eruption is because the paintings and illustrations that relate to the concept is not consistently showing the year 2012, but some Judgment scenes can be interpreted as a year before or a year after. We shall keep the main date of 2012 in our focus, because that is what the very accurate Mayan Astronomy priests marked down for us.

Although, in the last few years we are in a Solar Minimum, by 2010 – 2011 the NASA scientists are expecting a much more active

Sun. Even just only five six years ago, on November 4, 2003, the Sweet Sixteenth birthday of Austin Gaspar the son of Dr. Gaspar, brought us an X-28 solar flare. It blew away in a different direction from the globe, but if it been Earth-directed it could have had already devastating consequences for our current civilization. Fortunately, it did not hit our globe, but it is concerning that neither the mainstream media nor governmental scientists would loudly popularize the event as a warning for future eruptions to threaten the survival of humanity. The second largest sunspot eruption happened days before that one, on October 26, 2003. This initial eruption was X-17. This was matched again just in recent years on September 7, 2005. Gradually, the years 2010, 2011 and 2012 will begin again to demonstrate a more active Sun. The Sun can become so active, that by the end of 2012 we shall be ready to see a Huge Sun spot eruption.

Are we getting ready for the Big One in the near future of our current civilization? Right now, we are in a period of minimum activity, because the Sun is not having too many sunspots, but that does not

Figure 12-2: The Sacred Cosmic Union is depicted on this Erotic Art From the Temple of Kandariya Mahadeva, Khajuraho, Madhya Pradesh, India, circa 1050 CE. (Photo Credit: © DeA Picture Library / Art Resource, NY)

The World Tree 291

mean we cannot have giant coronal mass ejections, or so called CMEs, all of a sudden. These eruptions in the past may have been equated to the 'eruption of a male god'.

This temple façade depiction from India suggests the Sacred Cosmic Marriage. The cosmic female is standing and the male is upside down. Certainly, this is one of the versions of the two sexless forces of the Sky and Earth uniting. Generally, we think of our globe as a female sex, but the Egyptians commonly depicted Father Earth as Geb, and the sky was Nut, the female principle. Naturally, amongst the vast examples one can find all around the world, including Egypt, both a female and the male Earth or invading upper force is imaginable.

On the temple art from Kandariya Mahadeva, India the two females on either side of the mating couple seem to symbolize the Two Pillars of Mother Earth and also the Two Venuses of the Venus Transit Pair. The outside females represent a 9 division that ties Venus to the Sun, and the inner females are the 7 divisions relating to the cycles of

Figure 12-3: Amun-Rá is in the Solar Bark. From the Tomb of Seti I, Valley of the Kings, Thebes, Egypt. (Photo Credit: © DeA Picture Library / Art Resource, NY)

Mother Earth and her Pillars.

Thus, the 36 years square with 9 divisions on each of the four sides, likely differentiates between the Two Females identified by the nine divisions standing for the Two Pillar controlled by the Sun, and the Two Females standing inside separated by seven divisions and representing the seven years between the Venus Transit Pair. With that in mind, we shall examine in the near future the Egyptian depictions to see if we have a similar set of divisions.

In the solar bark of Rá, Amun-Rá or even Khnem, the Ram or Goat headed deity surrounded by the two pillars and the brightness of the Sun, we can find that Rá here represents the LAMB who reaches across the end of the Fourth Hour of the Book of the Dead all the way into the Fifth Hour. As we study this Egyptian cosmological depiction, we can count out that this Amun-Rá is truly standing in the middle of the Boat as the Fourth person in the Solar Bark. Hathor, the third person in the solar bark and the one right before Amun-Rá wears the Horns of the Celestial Cow with the Sun in between the horns.

Without the obvious Dot in the center of the Sun, this Opera sign can only truly represent, either the date June 8, 2004 or the later date of June 6, 2012. Certainly, by know we are assured that it represents the summer of 2012 and several months later we shall witness the power of the Sun in December of 2012 or 2013. When one even examines the body shapes of these deities standing in the boat, it is interesting to notice that above the belt their bodies display the deltoid or rhomboid shape that one would find inside the five pointed star, just as the Keystone body of Hercules in Astronomy. Then under the belt we find the mirror image of that same shape pointing to the opposite direction. This, and a few hundred other little details we have not mentioned, prove our points that the ancient hidden secrets were partly based on the concept of the two five pointed stars uniting. The black hair, bronze red skin, yellow skirt over the white tunic maintains the cosmic four directional colors we employed in our Cosmic Model. The **Rod** that Amun-Rá holds in his right hand already maintains the shape of the **Serpent** that we connected from the biblical stories of the Old Testament.

Our dilemma between the two years that would serve as the FOURTH Hour of the Last Twelve Hours of the Heroes Journey is

The World Tree 293

the fact that we do not know when the ancient year ended for the Egyptians. If it ended on December 21st, then the year of Hathor on the previous figure would end the Third Year on December 21st and we would find ourselves looking at the first few days of the Fifth Hour when the first eruption occurred, unless there was something already going on earlier in December of 2012, to qualify Hathor for the Fourth year. If the year ended on March 21st, then even with that it might be a little confusing. Thus, according to this picture from the Tomb of Seti I, in the Valley of the Kings, Thebes and another depiction from The Book of the Dead, we are not 100 % certain whether the first Huge Illumination appears on the night of December 24 - 25th of 2012 or the year 2013? One should prepare for both possibilities.

Why are we scientifically so sure about the fact that the Mayans were right about the soon approaching End Days, whether it will culminate in a beautiful dimensional shift, a form of a Rapture, a rescue mission from Space Aliens, the return of the Messiah, the return of Quetzalcoatl, the second coming of Jesus Christ or just plain earth changes?

Our argument for that is the follows. In the above figure titled Rá in solar bark, we can see that the obvious representation of Hathor, who clearly stands for the Transit of the planet Venus in front of the Sun, when the Sun is in between the Horns of the Taurus Bull. Thus, Hathor is within a year or so of the now obvious period of three months of solar eruption and the Axis shift represented by the Lamb, Amun-Rá. Now, since we know from the Ice age cycles that these Half Precessional cycles suddenly switch about 11,600 to 11,700 years apart, plus minus may be 50 years or even less.

Therefore, when we search for another Venus Transit in front of the Sun we look back and realize that since these Venus Transits happen 121.5 years apart, thus the one that happened 121.5 years before or after the 2004 – 2012 Portal will not occur in June, but in December when the Sun will not show in between the Horns of the Bull. Thus, another date would be 243 years away ahead or behind of today's Portal. That 243 year period is largely outside the scientific period allowed for any variations. The fact that the Ancients marked down 96 pairs of 'Two Ladies' (the secret name of the Portal of the Venus Transits in Egypt, shown by the Serpent and the Vulture), which would calculate out

96 x 121.5 = 11,664 years, and knowing that the even pairs of Venus Transits would always fall on June and not on December reassures us that we are talking about the last few years before a major reversal in climate is due.

A paid governmental scientist might disagree with us officially, but the numbers talk for themselves. Should we trust the ancient Egyptian pictures, along with the accurate Mayan calendar to estimate when the next one will hit, or should we simply trust our scientists? We feel that the Mayan calendar is more accurate. The scientists whom we contacted kept arguing with us that we are not yet near a solar maximum, thus we don't have to worry about a sunspot eruption. Then, why - we ask, even on November 3rd, 2003 only One Day before the X-28 or larger eruption happened did the NASA scientist report that the 'storm is over'? Why were these same NASA and NOAA scientists on the edge of their seats after the eruption, not understanding what happened and why and would it happen soon again? Clearly, because they did not understand how a large eruption could have happened without a warning. From the size of the eruption they understood how serious it was, but there was almost no mention of it in the regular state media. The solar eruption was enormous! Examining the sunspots of October / November 2003, the solar wind speed increased from an average about 250 miles / second to over a 1,000 miles / second. The instruments that NASA possessed were only calibrated to a maximum of 1,000 miles / second, thus they maxed out. This was a result of an eruption that shot away from the direction of the Earth. What if, one of these days, it hits the Earth, as it has done it before thousands of years before? Will the eight-minute warning be enough?

Increasingly, more scientists and lay persons are jumping on the bandwagon of stating that air pollution is responsible for the observed global warming, that is because they see obvious and concerning earth changes. But these increased magnitude climate changes are happening, because we are sliding into this Galactic Alignment. Certainly we had global warming in the last few decades, but Nature magazine reported in 2007 that we had the same amount of warming on planet Mars. That fact points to a Solar or Galactic cause, rather than a global cause. According to the Ice Age graph, soon we will start cooling. Those who think that the extreme hot temperatures and the recent cold weather

The World Tree

is global warming that is reversible, they are scientifically wrong! What does the faraway Sun has to do with what we humans cook up on this Earth? Any governmental scientist who thinks it is reversible global warming of human activity and would like to tell us why the Sun is acting so crazy lately – according to NASA - is certainly welcome to educate us in their logic. Luckily, good education in US Universities and foreign and American Medical Institutes allowed us to learn about scientific principles and how to read any scientific chart they present us. May be these scientists can influence the majority of the population with their semantics, but increasingly more and more people start applying their God-given logic to reject the false prophets of science.

Any common sense individual understands by now that the Sun would not go into idleness or would not overheat from the pollution we put into our Earth's atmosphere. The Sun is too huge and too far from us to expect it to overheat from our tiny earth's atmospheric pollution. If there were no human activity on Earth, if there were no industry whatsoever, the cosmic Mayan calendar would still be accurate and it would foretell when the Earth's Axis marches in front of the Galactic Center. Polluting our air is very wrong for several different reasons. We immorally do pump a lot of CO_2 into our atmosphere. So much CO_2 that we have passed the upper limits by 40 percent, but strangely, the temperature of the Earth is cooling, the winters are harsher and the Earth's Axis have not shifted, because it does the shifting on a very punctual cosmic calendar and not on human design. The government scientists, media and well wishing and caring individuals are fooled into believing that we are responsible for what is happening on our globe today. Please, do not pollute and do your recycling, because those are the proper things to do, but also please do not believe that what is happening to Earth is in our hands.

The Russian military has underground installations in the Ural Mountains and elsewhere, for the reasons we are talking about. Probably, nobody knows the exact timing of these solar hick ups, but they study the Sun and they are openly worried about the Sun acting up. Not only the Russians, but the Chinese, the European and also the US government is similarly preparing for just such events, but naturally they cannot house 300 million people if something awful happens. Well, the larger public will not be scared into a panic situation, they are cheerfully being told

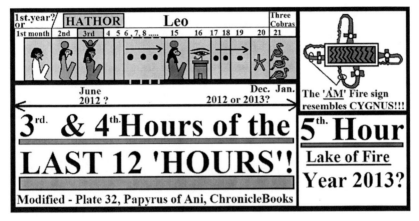

Figure 12-4: Hathor, Mistress of the West is depicted as the Third Person sitting in a Row of Gods in the Third and Fourth Hours of the Last 12 hours. This Hathor would represent to us the Venus Transit of June 5-6, 2012 and the Third Hour (Modified by Wm. Gaspar from Plate 32 of The Egyptian Book of the Dead, Chronicle Books)

about a bright and exciting future propelled by the Hybrid cars of the future. We beg to differ as one can learn from our dissertation.

According to us, the alignment with the Galactic Center will culminate between the years 2012 and 2020, and according to the Mayan and the Egyptian calendar, more specifically the first huge hits will follow one another in the years of 2012-2014. These years will fit the time frame for the beginning of the seven years of famine.

At the end of the Fourth Hour then the Sun Erupts – see right side of the depiction with the STAR followed by the Three Serpents. Since each god represents a month to us here then the Star sign would be DECEMBER 25th, 2013 and the Three Serpents stand for the future date of JANUARY 23-26, 2014. Unfortunately, other depictions also suggest an earlier version where by the end of 2011 we would experience serious solar directed weathers. Thus, our suggestion is that by the end of 2011 and the beginning of 2012, most everybody who is preparing should at least have a shelter underground. Water and food needs to be stored protected, because when the real huge fireworks start, the few weeks left to prepare is only serving those who are in place already and the warning shots would tell them to shop for some more.

Different information pointed at different times. This above picture was the one piece of information that told us that the December 21, 2012 date may be One Year later. Thus, if nothing happened by the end of December 2012, our advise is that please do not let your guards down, because the date may be correct, but only one year later. Unfortunately, nobody can tell the future, not even from the exact knowledge of the past, thus all we can do is to estimate and guess a possible earth change scenario that maybe too elusive.

Let us now rejoin the Egyptian festivities.

Hathor, who may represent the planet Venus or even Mother Earth, likely right side up or upside down with her legs spread apart as a harlot would do, was ceremonially carried to the roof of the Temple of the Goddess in Dendera at the change of seasons. The priests piously walked to the top of the Temple with the sacred bust of the goddess, to expose her to the rays of the Sun. This they did to mimic the unification of the goddess to the Sun disk. That and the name Tammuz in Babylonia points toward the Groom being the Sun King and the Harlot could only be Mother Earth, since she is the one who will be scorched by this intense cosmic love affair.

This was the SACRED COSMIC MARRIAGE of the male and female principles of Earth Changes. By the Syrian, Babylonian, and Talmudic Jewish wisdom calendars, the marriage of Tammuzi would be in the month of Tammuz and the 17th day. This, on the southern view, would compute out to the second week of July today, but on the northern view, which interests us the most it falls near the middle of March.

Hathor became Isis and her role as the married goddess was tied to the Sun and cosmic laws. According to professor Lesko, charms were made to honor Isis, to ensure that a woman would love a man as much as ISIS loved OSIRIS. Whoever made the charm must have not been aware of the burning cosmic fire and destruction, which accompanied that kind of celestial love. It was certainly a sizzling cosmic romance that most of us would love to live without.

The same kind of charm was employed if the man was after a married woman. If she were already married, then she would hate her husband as much as Isis hated Seth. The Knot of Isis was usually made out of red jasper, which symbolized the blood of Isis. This type of amulet was sent with mummies to the afterlife to gain protection from evil and harm.

According to the *Egyptian Book of the Dead*, by E. A. Wallis Budge, as recorded in the Paris Papyrus, Osiris is glorified in the following manner:

OSIRIS . . . you are established as the Bull of the West, your son HORUS is crowned on your throne, all life is with him. Your son, to him are given millions of years, the fear of him is for millions of year; fear him as the cycle of the gods.

Another spell in the papyrus of ANI instructs the reader through the transformation into a divine hawk:

Chapter of making the transformation into a divine hawk. OSIRIS ANI said: 'Hail mighty one, come then to TATTU (the two pillars). Arrange for me the ways make me go around my thrones. May I renew myself, may I become strong. Oh grant me your fearfulness, create for me your terror. May the gods of the underworld fear me in their habitation they are for me, Let not the One come near me that would do harm to me, may I walk through the house of darkness.

An interesting function of the Goddess ISIS was her search to reveal the divine name of her father, the Sun King Rá. According to legend, the old king Rá was having a walk one night and, because he was so aged and senile, he was drooling badly from his mouth. The saliva from this drooling fell into the earth. ISIS scooped it up and mixed it with earth and fashioned a snake out of the resulting mud. The snake then bit the old king Ra. (This is very similar to the way Greek hero, Hercules, got bit by the astronomical Cancer or in other tales by a Scorpion). Rá, the Old Sun King, begged his daughter to use her magic to get relief from the misery of the snakebite, but Isis remained stern. She would only attempt to use her magic for his benefit if Ra would reveal his secret name to her.

Finally, Rá could not stand the agony and pain, and revealed his secret name to her. From that moment on, ISIS obtained enormous powers from knowing the name of the sun king, Rá. This name is unknown

The World Tree

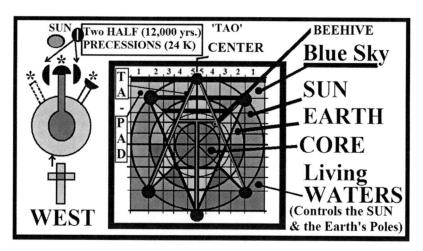

Figure 12-5: The main Form of Alignment (Drawing by Wm. Gaspar)

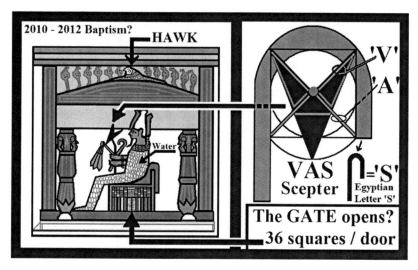

Figure 12-6: The meaning of the word SA-TA and SA-TA-N. As we recall the 'SA' is the Goose, the controlling Force from the Black Hole and the 'TA' is the Belt of Orion, where the Axis shift begins. The letter 'N' is resonation. As we demonstrated earlier the 'TA' syllable of SA-N-TA Maria is also incorporated into the Nativity Scenes as part of the Roof above the Sacred Birth. (Drawing by Wm. Gaspar)

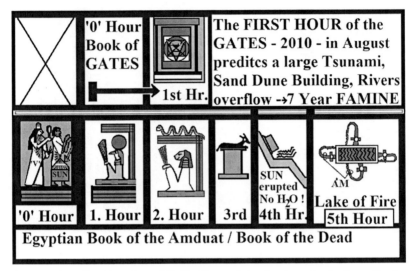

Figure 12-7: The Fourth and Fifth Hour of the Amduat in the Last Twelve Hour of the Egyptian Book of the Dead (Graph by Wm. Gaspar)

because it is not a name, but it is the resonation which comes out of the mouth of the Sun God and causes some major earth changes.

The hieroglyphic word for the evil Satan begins with the same cosmic Goose, which main star is X-3, the strongest gamma source from our galaxy, as the name for the purifier Son of the Sun. Although we read this name as SA-TA from the hieroglyphic symbol of the Goose (SA), and the belt (TA), there is one more figure there: the Serpent. If we equate the Serpent with the wavy line of resonation (N), which the Serpent actually stands for, then the full name of this picture would read SATAN. The Goose / Cygnus, Orion, and the Serpent Resonation are all integral parts of the cosmic concept of the "Coming of the Son of the Sun, who is the fiery Purifier of sins." It is possible that the silly satanic cults and their more serious opposing fundamentalist religious groups are actually keeping alive the legend of the same cosmic tragedy without knowing?

The Sun's role is not mentioned in that hieroglyphic picture. But as we can see now, the Egyptian depictions commonly show the Serpent in the middle of the Sun. The Judeo-Christian Satan is usually shown as GOAT-like human, but it is not an opposing idea, since during that alignment is when the first larger eruptions will happen as the birth of the little goats.

A quote from Erik Hornung's magnificent book, The Ancient Egyptian Books of the Afterlife, mentions snakes in plural in association with the Sun eruption when talking about the Fourth Hour, of the Last Twelve hours of the AMDUAT as we demonstrated earlier in this chapter.

Thus, as we have learnt earlier, the Fourth Hour ends the abundance of water for the last twelve hours of the Amduat. We don't know when this is going to happen, but we think it will be around the year 2012, especially if the famous Year 2008 is the Beginning Year Zero for the Last Twelve Years.

Then, the Fifth Hour of the Amduat brought the Fiery Lake. This may be the Year 2013 or 2014! What would allow a fiery lake to develop here on Earth? The Lake of Fire is tied to the Sacred Door, as the Sleigh that is sliding down into an unforgiving hellish state. H. A. GUERBER writes about Vulcan, the God of Fire, in the book *The Myths of Greece and Rome:*

> *VULCAN, or HEPHAESTUS son of Jupiter and Juno, God of Fire and the forge, seldom joined the general counsel of the gods. His aversion to Olympus was of old standing . . . The intervening space between heaven and earth was so great, that VULCAN'S fall lasted during one whole day and night . . . he injured one of his legs . . . left him lame . . . VULCAN . . . withdrew to the solitudes of Mount Aetna, where he established a great forge in the heart of the mountain, in partnership with the CYCLOPES, who helped him manufacture . . . objects from the metals found in the bosom of the earth.*

In the Lake of Fire there are dead bodies or evil spirits floating on the bottom of the pool. The fire is putting out fumes that may represent the plumes of the erupting volcanoes. But we have a Lake of Fire in the Bible, also:

> *"And Fire came down from God out of heaven and devoured them. The devil that deceived them was cast into the* **lake of fire** *.*
> —Bible, Revelation 20:9-10

Thus, in the above quote the Fire from heaven may represent the Fourth hour of the Egyptian Amduat and the lake of fire is the Fifth hour. It is not at all clear or appears as a continuous journey in the Bible. The fact, that the Fire everybody associated only with Hell is coming straight from God and right out of heaven is a little confusing, but clearly supports our idea of a Sun eruption! We know scientifically that it is how things happen in our earth change scenario. It foils the idea of an idyllic and serene heaven one can descend to after a well-deserved death. There are no safe places, anymore. Heaven is as fiery as Hell should be. The Devil or Satan with his fire from Hell seems to have a trick all the way up to heaven. Maybe the Day of the Lord is not different from the hellish fire and brimstone, whether it comes from above or from below. Very likely it is all the same. Purification can only proceed with almost total destruction. Now, one can say that this boiling heart has nothing to do with the shift of the Magnetic North, but our accurate Holy Bible states differently:

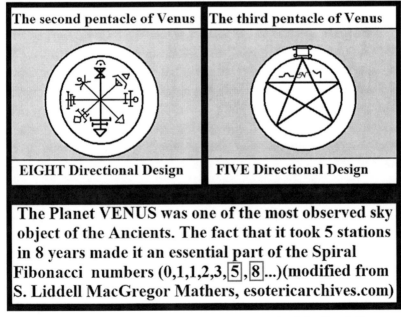

Figure 12-8: The Lake of Fire according to the Book of the Dead (Drawing by Wm. Gaspar)

Figure 12-9: Tata Alejandro and Pepe Jaramillo.

Then the Lord said to me: "You have seen well, for I am ready to perform my word." And the word of the Lord came to me the second time, saying: "What do you see?" And I said: "I see a boiling pot, and it is facing away from the north.

—Bible, Jeremiah 1:12-13

The "boiling pot" that is "facing away from the north" is the Magnetic North shifting toward the south direction to the location of the Polar North, according to our theory. He will "perform his word;" thus, it is a serious act rather than casual conversation! So, this boiling pot leaning away from the north will be caused by the "word" of God that comes as a strong resonation out of his mouth. How good it is to understand the scripture scientifically for the first time, when we now know that, in Egyptian the letter R is an open mouth and the letter A is an arm in a 90 degree angle, together the word Ra, the purifier.

The Egyptian Lake of Fire is found in the papyrus of Egypt. Did the Lake of Fire designation describe an event such as the eruption of the Super Volcano of the Yellowstone Park, or was the boiling inferno associated with the Ring of Fire of the Pacific Rim? We may soon witness it again, as we are nearing to the end of an 11,700-year old era when these ancient tragedies can repeat themselves. Will we be witnesses to incredible powers moving and shaking up our resilient Mother Earth? Nobody knows for sure, but the evidence has started piling up toward an impending natural cataclysm.

We would like to end this chapter with a quote from Tata Alejandro Cirilo Perez Oxlaj, leader of the Mayan Elders from Guatemala regarding the prophecy of the Eagle and the Condor:

I had the vision to do these gatherings; and so, in 1995, with the help of the Creator, I organized the First of these gatherings here in Guatemala, in the Center, the Land of the Quetzal the Land of the Mayans. Like our Mayan Prophecy says: "Those of the Center may unite the Eagle of the North with the Condor of the South, we will meet, be\cause we are one like the fingers of the hand." Two years later, in 1997, the Second gathering of Indigenous Elders was held in South America, in Bogota in the Amazon jungle in Colombia, the Land of the Condor. Two years later, in 1999, the Third one in North America, New Mexico, USA, the land of the Eagle. We joined Americas! After the Third gathering, the Sacred Staff that had helped us join the three Americas was gone and the gatherings stopped. Eight years of stagnation followed. But now the Sacred Staff is back, the movement has returned. Now we want to reach all corners of the world. All those who come to participate in this gathering will be Messengers of the Prophesies, Messengers of Peace, Warriors to save our Mother Earth and all those who inhabit her: brother animals, trees, all living beings and the world of the stone people. Together we can do it. Together we can bring peace.
—Tata Alejandro Perez Oxlaj

Thank you father for having given us life, bless us all those who are here on Earth, thank you Father and please bless our people,

our mountains, our rivers, our lakes, all that we are and thank you Father." You and I may meet again in another dimension after the year Zero. The year Zero is the word of the Maya. On March 31st, 2013, the sun will be hidden for a period of 60-70 hours and this is when we shall enter the period of the Fifth Sun.
—Don Alejandro, leader of the Mayan Elders

CHAPTER **13**

The Real Cosmic Secrets and The Da Vinci Code

To become a real boy, you must prove yourself brave, truthful, and unselfish.

—Pinocchio

We couldn't finish this work without talking about Dan Brown's book and also use the occasion to address some spiritual and survival issues.

The secret behind the novel of Dan Brown's, *The Da Vinci Code*, is phenomenal and, to some extent, unexplained. It has concoctions of conspiracy theories, Illuminati world control, and spirituality, and it plays on some strong religious beliefs. First of all, let us share our belief that everybody on this planet is the perfect product of the creator force. We don't always act it or acknowledge it, but we have a Spirit / Soul that was provided free of charge from our Creator. That conscious Spirit, if allowed by the mind and the flesh, can make—in most instances—good, rational decisions. There are NO humans whom we could label rejects, that we could place a sticker on for everybody to see that said: "*Look! This human is a fake. He was made with natural and ARTIFICIAL ingredients!*"

None of us is perfect, and we are not at all genetically sound, but spiritually, we are from the same godly resonation. Sometimes the most pious people are the ones who read their holy books the least, or

The World Tree 307

if they do read them well, they then seem to favor certain parts and skip others. When we read any holy texts, including the Torah, the Rig Veda, the Koran, or the Bible, to us they clearly state that God is a SPIRIT, and we will never see his face, and we are ALL his perfect spiritual creations.

Did he produce faulty stuff initially, to later correct it, and then have his son marry a mortal woman who slept around with multitudes? Can, somehow, the product of that union be the holiest or holier than those who try very hard? We don't think so! We hope that this scenario seems crazy both to the spiritual and even to the religious people.

We believe that we are all perfect little products of the Perfect Almighty Creator who gave us our Spirits from Its Spirit. If that Spirit Force impregnates the whole universe and we are spiritually connected to IT, then we don't see how we need a special race or a special daughter or grandchildren of a special son of God. Aren't we all his & her special children from all walks of lives and all denominations? Sending his son to teach morality and spirituality may make perfect sense to those who believe that we need to be reminded to be spiritual by him because our vain desires control our lives. Yes, that may make perfect sense. From any direction, from any denomination, we humans sometimes forget that the Spirit leads and negotiates the mind and the body. If either the mind – in the case of serial killers and religious zealots - or the flesh body takes unreasonable control over our lives, then we know that we are in trouble. No question about that.

Having a mutant, half-human half-god creature has been advocated in mythology from the Babylonians to the Greeks, and everywhere else. It did work for them only as weird mythology, but not necessarily as history or moral and religious teachings! This half human half god flesh being should not work for Dan Brown as history or redemption of any source. We shall leave it at that.

So, when we speak to people of possible serious Earth Changes in the coming years, people often ask us: Aren't you afraid that we all going to die?

No, we are not afraid. No, we are not afraid to die, if that is the question. The first wave goes fast and innocently. The rest . . . who knows? A lot of people will survive. They always did in the past. Dying is the lot of every human being, sooner or later. Dying with dignity in any situation is a good death. We are amazed by elders who can one day say, "Well, son, I think this is my last day here on earth. Let me go out to the garden to water the tomatoes one more time." Then they put on their best clothes and next morning, they don't wake up. These people have to have a strong belief in their Creator that he or she will take care of them on the other side, and why not? If the Creator can make order out of hundreds of billions of galaxies traveling around in the endless universe, then he or she probably figured out what to do with our souls after they depart from the flesh. That is not the problem, and should not be the source of fear. In our believe death is only another beginning, another gift of the Creator for us to exist in a different form.

We do have some real fears and apprehension about a sudden earth change scenario when large masses of people who are the unwilling, unprepared, and unlucky – will survive along side of those few who prepared. The unprepared, that is the large majority, will have to make some hard choices. That is the time when a person with love in his heart toward his Creator will face some tough decisions. That is the time when we will feel fear that we have not prepared enough, and that we have not warned those who will be left behind, that we may have not used the right words or the right examples to wake up the cautious sides of their spirits. At that point, the Spirit of the unprepared will not be able to guide the mind and the body to a prosperous life, since by then there will be no resources widely available for years to the majority of the population to exist in a decent human style. The survivors might have to resort to be the villains or the victims of unusual practices, such as the unpleasant diet of cannibalism, killing, raping, looting and all the evil deeds they tried to avoid in their decent lives.

We don't know if there is any special, saving-your-soul method that we could employ. Sorry, but everybody is on their own in their own belief system on that chance. Is there karma attached to our actions in the last hours? Is there a dimensional shift, a good will alien involvement to evacuate the Earth or a biblical rapture? We don't know. We hope there is, but the human reality is that natural disasters happen and the saving magic associated with it is seldom there. When these physical realities become increasingly obvious that we may have a major one coming soon, it behooves us to have some form of spiritual, mental, and physical preparations. That's all we can say about that. We wrote the book to explore the worst of the possible climate shifts that could realistically happen according to what is said in world mythology. Will it happen that way in a certain year? We can't say! The question of how to survive a major earth change scenario is also not in the scope of this work.

Let us return to The *Da Vinci Code*. The manner in which they disclosed secrets in the book is an old trick of the ancient sages. Anytime there is a secret, then the story is made enigmatic by several means. The story is almost always the same in the Mono-myth, where the uniting and cosmic forces causing catastrophic earth changes as the gods interacting with humans or, even more specifically, as the marriage of *The* Divine Couple and the Birth of the Divine Child.

In the following, short essay, we shall demonstrate that, to unravel the secrets of *The Da Vinci Code*, one needs to look at astronomy, cosmology, and the recently discovered workings of the Earth's geo-dynamo.

The book begins with a powerful image of the leader of the Secret Society being murdered to give up his secrets. He would not, he would rather die. He assumes the position of the Five pointed star in his death. We spent a good portion of our research to demonstrate the cosmological meaning of the Red Star as the Magnetic Axis shift. Since, we know that the Freemasons cherish the Compass, and we have witnessed the Grand Architect of the Universe measure the Earth, we concluded the ancient secret about the Compass is the fact that it represents the knowledge of the Two Pillars of the globe. The Square we explained as the 90 degree stabilization of the Sun on the Magnetic North. The letter 'G' stands for God and Geometry, where the letter making secret of the ancient Phoenicians, Egyptians and Romans were found to be hidden in the same five pointed star. Thus, our five pointed star, or the Pentagram that represented the Full Precession of the Equinoxes, the Red Hero, Hercules and also the Harlot, Mother Earth and furthermore the hieroglyphic picture letters of the Egyptians could be an important secret. None of these ideas were explored in Dan Brown's book.

We have demonstrated a number of secrets of the Pentagram. How much more one can find in the simple form of a five pointed star that faithfully was preserved on the flags of multitude of countries today? The five periods of the Heart beat of a human compared to the five waves of the Ice ages and the five waves of 4.8 hours that turn the Earth from west to east. The five stations of the planet Venus around the Sun, the five divisions of the Half Precession of the Equinoxes and much more! How much more sacred things can be found out from the simple shape of the FIVE POINTED STAR? Thus, if the knower of the secrets, the leader of the Secret Society who was not afraid to die for his belief, truly wanted to show something sacred, something as enormous as the power of half of our galaxy, heart beats and all, then he would have shown us the secret knowledge about the Star that we presented throughout this book. That is worthy of a noble death if that is the reward!

The Da Vinci Code describes the Rose Line as an invisible straight line, drawn from the North Pole to the South Pole that represents the original prime meridian (0 degrees longitude, before it was chosen to run through Greenwich, England). According to Dan Brown, the Rose Line passed right through the Church of Saint Sulpice in Paris, along the brass line on its floor. Actually, that line passed by this church over **one hundred meters** to the **east.** Maybe that was by plan, to mark the 100 leagues that the Real Rose Line runs between the Magnetic North and Polar North.

The term *Rose Line* was popularized by Dan Brown in his novel *The Da Vinci Code* as an alternate name for "the world's first prime meridian," identified as the Paris Meridian. Brown's novel also conflates this meridian with a gnomon in the Parisian church of Saint Sulpice marked in the floor with a brass line, as did the 1967 Priory Document, *Le Serpent Rouge, (Notes sur Saint-Germain-des-Près et Saint Sulpice de Pari)s* attributed to Pierre Feugère, Louis Saint-Maxent, and Gaston de Koker, who called the Rose Line the **"Red Serpent."** The Red Serpent sounds better when one knows that the Dragon sits on the Red Line between the Poles of Mother Earth.

In our World Tree theory, the **Red Rose**, or Magnetic North, is connected with the **keystone** and the core / foundation stone of the geo-dynamo of the Earth. The serpent is commonly used as the symbol for the Resonation out of the Sun in ancient times, thus we can observe the **SERPENT** wrapped around the Sun on the Head of

The World Tree 311

Horus and other Egyptian deities.

It would be a grave mistake to search for an actual **GRAIL,** or actual historical figures in the sacred tales of advanced cosmic knowledge. The ONLY GRAIL that will meet our astronomical expectations will be the GRAIL on the south side of the Earth's Axis shift. The ancients used two markers for the Red Road. One of them is between the Two North Poles where we find the Dragon and the second one is the extension of the original first running all the way down to the Ecliptic where the path of the Sun is and where the Cancer constellation is found to intersect the Serpent Hydra. Thus, we have two separate views for the same shift: one through the Dragon and one through the Serpent.

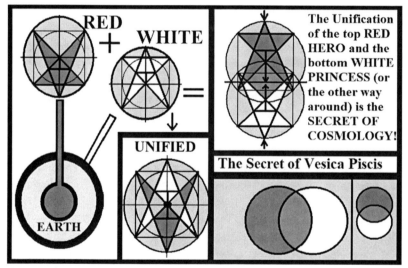

Figure 13-1: The Rose Line, or the Red Road, leads through the Magnetic North to the Polar North through the Dragon and through the Serpent (Graph by William Gaspar

Once one is aware of the astronomy of these areas and what time they would appear on the Zodiac of Dendera, then one can tell the timing of events by knowing the animal symbols.

Religious leaders and their followers throughout the ages have propagated a story line that did not give an explanation of cosmology, only historical sounding moralized animal tales of questionable significance. This was the way to maintain very important information about the timing and the mechanism of the severe Earth changes that happened at the beginning of the Age of Leo about 12,000 years ago, and the likely occurrence of similar Earth axis shift at the end of the Great Mayan calendar in 2012 or 2013.

We do not have the luxury to play games and speak in the form of riddles anymore. Therefore, let us forget about the old trickery of "transforming the Divine into Human," as that is only the tale that carries the sad, but sacred message of unavoidable periodic Earth destruction. We now need to translate those stories backwards and find out what cosmology is hiding for us. We are certain that neither the talented Dan Brown nor his extremely clever editor believed in the marvelous ending they manufactured for their fictional story. They applied the old trick of **humanizing the divine** to be able to deliver some important secrets to the public that would unlock the first three doors of the last seven on a corridor of advanced cosmic knowledge. This hallway leads to the unwrapping and opening of the Pandora's Box of secrets. The worldwide success of *The Da Vinci Code*, both literary and cinematic, proved their clever method right. Thus, we are not writing this essay to demean the artists who brought us this fictional masterpiece; rather, we applaud them. Their bravery in keeping open the first few doors of cosmic secrets in a largely dogmatic and ignorant world allows the rest of us to bang on the few last doors, leading to the enigmatic interpretations of the secrets.

Unfortunately, the SECRET is the COSMIC KNOWLEDGE of the ancient ones, and the precise ASTRONOMICAL PATH leading to the periodic Earth destruction. It is bad news and maybe it is better kept a secret for those thousands of years! Since we are standing at the doorway of the next cosmic destruction of our peaceful earthly paradise, be it now or thirty years from now, this secret becomes crucial for some and annoying to the rest. Should we open up the box of secrets, or is it better not rock the boat? If one already figured out how to be "saved" by the numerous options that exist out there, then this book might be an unimportant exercise in futility.

Those who think that they might not fit into any life-saving boats of this sinking Titanic need to begin making some timely plans for the expected cosmic shift of spirit and flesh. We demonstrated throughout this book that the Holy Grail is an astronomical and a cosmological

The World Tree 313

concept of the Earth axis that shifts every about 11,700 years. Rather than being a physical object, we clearly depict it as the astronomical GRAIL (Crater) in the center of South Celestial Equator (Chamber of the South). When this axis shifted and turned the ocean's waters to blood color by the magic of the Birth of the Sacred SON OF THE SUN, this was apparently achieved by erupting underwater volcanism.

Most of us commonly refer to our globe as our **Mother Earth. This tiny dust particle is a sacred island in the endless waters of the Universe and it is our home.** If she is equated to Guadalupe or any form or teachings of the Virgin mother of the Sacred Child, including the Holy Grail, we are not offended by those comparisons of life maintaining forces being of our Mother Earth. For the conclusion of this book, we shall go back to the wisdom of the Hindus whose written cosmological teachings are the best resources of untainted mythology.

Three excellent Brahmin sons were born of the earth when she made love in the form of a female boar; ... These children played affectionately together in the caverns and lakes on the various levels of the tablelands of the golden mountain Meru. The boar, surrounded by these sons, made love with his wife and did not consider giving up his body. ... 'The earth is crushed by the constant play of the sacrificial boar' said the gods. 'All the worlds are shaken and find no peace'. ... The powerful boars have broken the trees of the gods, the coral trees in Indra's paradise ...they made such a deluge as they fell into the ocean of salt that they stirred up inundations of water that flooded the whole earth.

–Hindu Myths by Penguin Classic

Thus, as we look at the next paintings from India that displays the marriage ceremony of Vasudeva and Devaki, the parents of Krishna, who as much the Son of the Sun as the others after him. His cosmic birth will be the prelude to the erotic love play of the cosmic boar with the Mother Earth, who will be represented by the parents of Krishna. It is quite understandable that the myth makers of any culture or on any continent, did not desire to place the picture of a Cosmic Boar making love to Mother Earth, as the parents of the Son of the Sun who was touted to be the Savior or the Redeemer. That would not paint a Holy Picture of those we like to pray to, although cosmologically it would be closer to the truth.

The secret of all ages is always cosmology. The Pentagon shape is present in Astronomy and it represents the Spiral with 5 divisions, thus it is the Half of the Whole. When the Half Precession of the Equinoxes are reached every about 12,000 years then the Sacred Cosmic Ceremony of the Celestial forces has to be performed again. That is what the Marriage Ceremony of Vasudeva and Devaki, the parents of Krishna accomplishes for us.

Figure 13-2: The Marriage Ceremony of Vasudeva ('VA-S ...VA') and Devaki ('..VA..'), the parents of Krishna, circa 1760. From a collection of illustrations to the 'Bhagavata urana' Location: British Library, London, Great Britain. (Photo Credit: Erich Lessing / Art Resource, NY)

The top of the marriage tent clearly ends in a yellow five pointed geometrical shape. The Yellow or the Red Star is where cosmology begins and ends. It is not the intellectual possession of the Communist Soviet Union, or China, North Korea, Turkey, USA, Texas or Arizona. It is also not the sole trade mark of Macy's, Heinekens, or even the Secret Societies.

The World Tree

This human shaped star is simply the cosmological secret possession of all human beings, beginning from before the time of Atlantis and before the last Huge Flood until now and for the unseen future.

The Cosmological or Astronomical Priesthood of all ages saved the same wisdom from every corner of the world as part of the universal Mono-Myth. It is important to understand, that if we desire to know the SHIFT to the next NEW WORLD ORDER, it is obviously NOT directed by secret societies or corrupt priesthoods of any kind, but, rather, brought on by the Great Architect of the Universe! No secrets; only an unavoidable faith of transformation!

Are there conspiracies in this world? Well, all we can say that, when we feed our horses, dogs or cats, then we usually see the lead animal push away the other ones and then the leader eats the most. Sometimes, the dominant alpha male will let the second most dominant in line eat there, too. So, is there favoritism among different groups? Certainly, there is. It is present in the natural order of things, beginning with one-celled organisms all the way to the humans. When it becomes uncontrollable greed that effects the survival of the group of animals or humanity then it has to be stopped. Thus, we should not spend too much time figuring out who is in a secret society or who is not, especially when our popes are changing the religious calendars to match the ancient Egyptian calendar of the Zodiac of Dendera. No need to find the most secretive person when our own Son of God is a Freemason and is depicted on a thirteenth century French manuscript with one of the telling Freemasonic tools, the COMPASS.

The only secrets, we average people don't know are the ones that we conveniently keep from ourselves. The vast amount of knowledge, that is now available to all of us on the Internet is mind boggling, and the fact that more people visit triple X sites on it than web pages that would educate and lead us toward a more enlightened state is not only the fault of our education system. We individual people are also responsible for our own fate and our own actions.

The sins of the fathers will affect us all, but the beauty of our spiritual existence is that if we pray or meditate for wisdom, keeping only the desire for pure truth as attainment, then every moment of our lives, we can restart new, be the best and seek for the true ways. Not one religion or one dogma, but any religious or spiritual way that leads

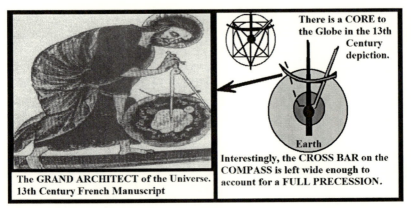

Figure 13-4: The Grand Architect of the universe. A French manuscript of the thirteenth century (Courtesy of Acharya S., The Greatest Story Ever Sold: Adventures Unlimited Press, Kempton, Illinois.)

to a true understanding of Nature and the "light."

Was Jesus Christ the last of the Egyptian Adepts of the Astronomy Mystery School? Was Krishna, Mithras, Mohamed and even Jesus Christ only a flesh manifestation of the mythological renderings of the Earth changes? We will never know for sure.

Our personal belief is that there was a reason why only highly educated astronomy priests could take on this knowledge. The Redeemers of different faiths brought us a cosmic knowledge that was needed to be shared with everybody who could finally understand it.

Religion and spirituality are there to teach us morality, but the religious and spiritual symbols of enlightenment are there to lead us to the one common truth, that the universe is ONE and we are one miniature part of that One. Therefore, an understanding our role and our endless faith is just to harmonize our souls and spirits to the One Universal Soul. Our role is to understand and accept the natural rhythms that are some ripples in the big oceans, and, sometimes, huge waves come, not to eliminate us, but to drift us into a New Age. That is what we think every religious and spiritual teacher is propelling us to understand and accept. Only our Ego, that is part of the One Big Ego can lead us out of Chaos, but that same Ego of ours can keep us isolated from the One and allow us despair. Our purpose is to find our

The World Tree

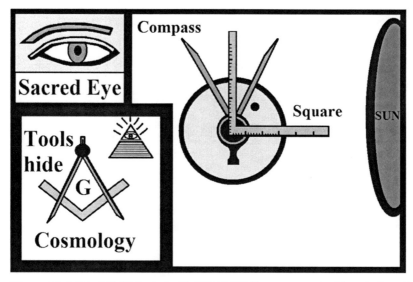

Figure 13-5: The TOOLS of the FREEMASONS represent the Knowledge of the workings of the Geo-dynamo in the Solar System and in the Galaxy! (Drawing by Wm. Gaspar)

resonation to fit the great, big Resonation.

Before we go into detail of the Rose Line, we want to mention that the most sacred flower for the people of Mesoamerica, specially the Aztecs and the Mayans, was the rose. This beautiful flower carries the same sacred symbols as the lotus flower, and as matter of fact, the rose was connected with the flower of life. That is why, in Mexico, as in other places, people buy roses for their loved ones. Definitely, the Latino culture is keeping the tradition without knowing the real, cosmic meaning of the hidden message behind the rose: the flower of burning love.

Also, in the Latino community of Mexico, a red rose is the sign of (cosmic) love, because it is the color of the heart. But the old tradition says that if you give someone a white rose, it is because somebody related to that person has died. It is very interesting to see how the Mexicans kept the tradition of the battle of the two roses by the meaning of the colors of the roses. One gives the red rose when you love someone, but then one gives the white rose for sympathy with mourning. It is also noteworthy that, if you mix these two colors together, you have

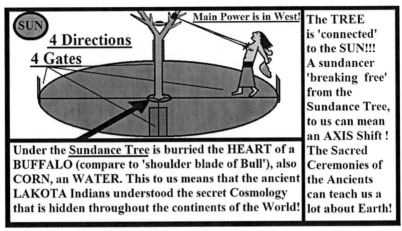

Figure 13-6: The Sun dancing Lakota wears a Red skirt and breaks away from the Sun dance Tree as the Magnetic North breaks away from its position. The Lakota Sioux Indians called the Spiritual Path the RED ROAD (Drawing by Wm. Gaspar)

the pink color that, in Spanish, is called ROSA, and when you give somebody a pink rose, it is a symbol of appreciation. Thus, when the two colors mix and become united as one, it becomes pink, the color of respect. Maybe respect is paid over the cosmic powers that mixed the red and white together.

This Line between noon and one o'clock was the Path of the Red Magnetic North to the White Polar North. Thus, this was the mysterious ROSE LINE, otherwise called the RED ROAD. This Red Road is clearly displayed in their Sun Dance Ceremony that helped us solve the Cosmological Mysteries. We are not ashamed that we learned Universal Cosmology from our people, the ancient Lakota and the amazing Mayans. We are very proud to be part of their mysteries and thank them from the bottom of our hearts that they found us worthy to include us in their Sacred Ceremonies. Without naming them, a number of medicine men and women passed our spiritual paths and shared their knowledge with us that allowed the development of this book. We feel that the Sacred Hoop is completed. The Cosmic Wisdom travelled across our globe and not one nation has to apologize for the parts of the puzzle their carried. Directly or indirectly, we have learnt

The World Tree 319

from Priests, Rabbis, Buddhists, Hindus, New Age gurus, Shamans, Medicine Men and Medicine Women.

After the symbols of the Freemasons and the awesome Hindu wisdom, we would like to complete our picture gallery with three more pieces to conclude our book. The first one carries the message of the Messiah and the birth of the Redeemer from the old and new Judeo-Christian point of view. In it again, we will recognize the universal Sacred Cosmic Colors and the secret TA of the Sacred Feminine of the Mother Earth as she proudly displays the birth of the Son of the Sun. The rays of the Sun are usually displayed behind the heads of the Holy Baby, the Holy parents, the priests and the angels. By know, we understand the specific role the Sun plays in these Holy matters.

On the last several pictures of this last chapter we demonstrated part of the Cosmic Wisdom of the Freemasonic Tools and their meanings then we looked at the Hindu and the North American Indian Cosmic Wisdom. To stay true to our callings, we will present three more pictures each representing a different belief and another part of the world displaying the same universal cosmic messages. Thus, our last chapter is about a combined Freemasonic Cosmology and the pictures from the 1, Hindu 2, Native American Indian 3, Judeo-Christian 4, pre-Christian Mithraism and 5, Babylonian mythological schools. Since, we have touched on largely Egyptian, Mayan and Hebrew Cosmic Wisdom in most part of our book, we felt appropriate to complete our book with other important international collection of religious and mythological gems.

On the next picture we will demonstrate the astronomical secret of the pre-Christian Mithras Bull Killing and the Twins, still an active part of most religious believes around the world.

Finally, we shall introduce a very special Babylonian Harp with a Bull head that will tie the Cosmological Wisdom of the Ages, including the secret of the Half Precessional shift.

With these last several figures in our last chapter we attempted to stay true to our belief's, that the Cosmic wisdom is universal and it is the possession of all human races of all believes from all part of the World.

First, we would like to draw attention to the Roof again that displays the famous TA of the Sacred Feminine. As the old barn is

Figure 13-7: Nativity with the FOUR Cosmic Colors of Black, White, Yellow, and Red. Painting by Rogier Weyden titled 'Nativity with the donor Pieter Bladelin'. It is the Center Panel of the Middleburg Altarpiece, circa 1445. Location: Gemaeldegalerie, Staatliche Museen zu Berlin, Berlin, Germany. (Photo Credit: Bildarchiv Preussischer Kulturbesitz / Art Resource, NY

disintegrating, we have the promise of a brand new temple in the background. The Sacred Cosmic colors of Black, Red, Yellow and White along with the Blue and Green are cleverly applied in this painting of Weyden. From left to right, first we can notice Joseph in a Red Robe. Then passed the first pillar a small angel dressed in Yellow is kneeling and praying over the Baby Jesus. Right above her and thus above Jesus is the always present face of the Sacred Bull or Cow, who to us in the Mayan Calendar scenario astronomically represents January 1st, 2013. The Virgin Mary is wearing a White dress, thus fulfills the requirements for the Sacred Cosmic Marriage colors in between the

The World Tree 321

Red Male (Magnetic North) and the White Female (Polar North). Virgin Mary has the Sun brightly shining behind her hair. She is kneeling on a Blue Robe that represents the color of the sky and water as she shares the robe with her holy son. The first Pope to the right is wearing the Black robe. It is interesting to notice that the Second Pillar of this dilapidated barn is missing. The Axis shift is imminent.

The obvious cosmic symbols and colors are astonishing. Who were in the possessions of the Cosmic Wisdom that we are only discovering? Who instructed these painters of the Dark Ages, when the rest of us comfortably existed on a flat Earth, to paint obvious knowledge of the Two Pillars of an active geo-dynamo on canvas? We feel cheated, as far as back as 1445 AD and even today. Where are You, Advanced Astronomy Priests of the Ages are hiding even today and why?

As we recall the head of the Secret Society from Dan Brown's *The Da Vinci Code,* we need to remember that he was a Curator at the famous Louvre Museum of Paris. We appreciate his fictional sacrifice, since this marble piece about Mithras had to be amongst the master pieces he defended with his life from the legendary Louvre of Paris, France. The killing of the Bull will remain an important aspect of ancient Cosmology. It likely did not impress the fine ladies of the Roman Empire, thus by the time Christianity rolled around, the Baby Jesus lays peacefully next to this Bovine astronomical symbol. The message is still the same, as Mithras birthday was on December 25th, too. Eight days later on January 1st, when the Hindu Male Boar first bumped into the desirable Mother Earth, the Hebrew boy had to be circumcised and the Mediterranean MIN-oan Bull had to die.

Under the belly of the Bull one can observe the Serpent. At the edge of the robe of Mithras we can find the Crab and above the CANCER are the statues of the TWINS. We can see lions and rams on this marble piece. More importantly, we detect Four Females on the upper side of the statue, just as we counted Four Females surrounding the Love Making scene on the Temple of Kandariya in India. Thus, we are certain that the Secret Formula of the Sacred Cosmic Marriage is Universal.

Thus, this is a pre-Christian statue of Mithras, called "Sol Invictus," the invincible Sun, whose birthday, interestingly, fell on December 25 and deserved the artistic merit to be displayed in Louvre!

322   William Gaspar and Jose Jaramillo

Figure 13-8: Mithras Immolating the Bull (from the Mithraueum of Sidon), Roman, 2nd – 3rd century CE. Photo: Herve Lewandowski. Location: Louvre, Paris, France (Photo Credit: Réunion des Musées Nationaux / Art Resource, NY)

The World Tree   323

One of the Egyptian Judgment Scenes. Pharaoh sits on the Throne (Cassiopeia) The NINE levels refer to Nine Years. The Baboon in the Solar Bark cuts off the Head of the BOAR. The Year here is not clearly marked with the Four Bulls.

Figure 13-9: The Babylonian Harp (Used by permission of the Trustees of the University Museum, Philadelphia)

The apparently astronomically-minded new priesthood of the greatest EMPIRE of the new era 2,000 years ago and had so much effect on our current civilization—the HOLY ROMAN Empire—was ready to remake the cosmic religion from the ashes of the old ones. Nothing stayed the same, but apparently, also nothing was altered as far as the cosmology of the new religion was concerned. The secret knowledge remained the same!

The same information of animal symbols and the sacred cosmic clock is present in Babylonia, where they unearthed a HARP that had the symbols of astronomy and the Half-Precessional switches marked on it.

On the top picture we notice the Hero between Two Bulls. The second picture from the top is even more exciting. The LION – whose Age represents the Half Precessional shift that happened 12,000 years ago – also is shown holding a KUP. This is the GRAIL the Mother Earth represents during the month when the Lion takes over the rulership at the time of the Axis shift. The Kup is also the letter 'K' of the Egyptian ABC and in the Sa-K-Ra-ment concept. In his left hand, the Lion hold the Vase that stands for the famed Age of Aquarius that for us began in the year 2008. These two symbols, the Lion and the Aquarian Vase represent the ages 12,000 years apart when the big shifts happens. Very innocent appearing animals 'going for some water' one would think, but by now in our advanced cosmological knowledge this little picture is a Huge Message! On that same level the Sirius Dog carries the Time

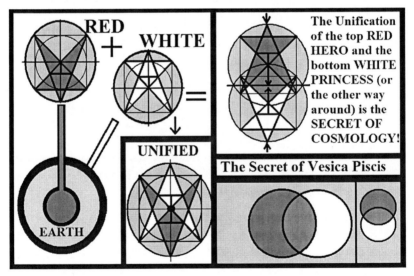

Figure 13-10: The RED STAR Unites Us ALL.

Piece, displaying the 'X' we find in the pyramid and the Secret formula. The head of a Lamb is next to the Leg of the Bull and likely a Wolf. All of the animals are precise astronomical markers of the earth changes. On the third level from the top we encounter a Bear (Polar North) who

The World Tree 325

is holding a Staff that is aligned by the Shoulder Blade of the Cow, while we see that the Donkey plays the Harp of the Magnetic North. All of these Cosmological Wisdoms tied into one kingly Harp decorated with a Bull Head! A worthy Art Appreciation class to conduct!

The RED STAR of Cosmology is the Major Secret representation of the legendary Heroes of the Creation stories and the Holy Books. Without the altruistic sacrifices of millions of spiritual people, priests, rabbis, imams, rishis, monks, shamans, medicine persons, artists, poets, painters and story tellers, who tirelessly rewrote, reworked, repainted and any which way rerecorded the legends of the ages, we would not have the numerous depictions to help us solve these enigmas. With our book, we attempted to honor all of those who walked before us, those who now walk with us and the ones who will walk the same uniting spiritual path in the future that we tried to follow presenting these tales. We are all related!

ΔΔΔΔΔ

BIBLIOGRAPHY

Acharya S, *The Christ Conspiracy*, Adventures Unlimited Press, Kempton, IL, 1999.

Allan, D.S., and Delair, J. B., *Cataclysm!* Bear & Co., Santa Fe, 1997.

Baltsan, Hayim, *Webster's Hebrew Dictionary*, Simon & Schuster, Inc., New York, 1992.

Barnstone, Willis, *The Other Bible*, Harper, San Francisco, 1984.

Bauval, Robert and Gilbert, Adrian, *The Orion Mystery*, Crown Publishers, New York, 1994.

Begg, Ean, *The Cult Of The Black Virgin*, Arkana, 1985.

Bell, Art and Strieber, Whitley, *The Coming Global Superstorm*, Pocket Books, New York, 1999.

Black, Jeremy and Green, Anthony, *Gods, Demons, and Symbols of Ancient Mesopotamia*, British Museum Press, 1992.

Blavatsky, H.P., *Isis Unveiled I & II*, Theosophical University Press, Pasadena, 1988.

Boyle, Ed and Weaver, Andrew, *Conveying Past Climates*, Nature, 3 November 1994.

Broecker, Wallace S., *Massive Iceberg Discharges As Triggers For Global Climate Change*, Nature, 1 December 1994.

Brown, Dan, *The Da Vinci Code*, Doubleday, Random House, New York, 2003.

Brown, Dan, *The Lost Symbols*, Doubleday / Random House, New York, 2009.

Budge, E.A. Wallis, *An Egyptian Hieroglyphic Dictionary*, Dover Publications, Inc., New York, 1978.

Budge, E.A. Wallis, *Legends of the Egyptian Gods*, Dover Publications, Inc., New York, 1994.

Budge, E.A. Wallis, *The Egyptian Book of the Dead*, Dover Publications, Inc., New York, 1967.

Campbell, Joseph, *The Hero With A Thousand Faces*, Bollingen Foundation, New York, 1949.

Cannarella, Deborah, *Christmas Treasures*, Barnes & Noble Books, New York, 2004.

Childress, David Hatcher, *Lost Cities series*, AUP, Kempton, IL.

Churchward, James, The Sacred *Symbols of Mu*, BE Books/ Brotherhood of Life, Inc., Albuquerque, 1995.

Churchward, James, *The Lost Continent of Mu*, BE Books, Albuquerque, N.M. 1998.

Clark, Rosemary, *Sacred Tradition in Ancient Egypt*, Llewellyn, St. Paul, MN, 2000.

Coe, Michael D, *The Maya*, Thames & Hudson, London, 2002.

Collins, Andrew, *The Cygnus Mystery*, Watkins Publishing, London, 2006.

Cooper, D. Jason, *Mithras: Mysteries and Initiation Rediscovered*, Samuel Weiser, York Beach, 1996.

Cotterell, Arthur, *Classical Mythology*, Lorenz Books, New York, 1999.

Cotterell, Arthur, *Norse Mythology*, Lorenz Books, New York, 2001.

Cotterell, Arthur and Storm, Rachel, *The Ultimate Encyclopedia of Mythology*, Hermes House, London, 2003.

Cotterell, Maurice M, *The Lost Tomb of Viracocha*, Headline Book Publishing, London, 2001.

Cotterell, Maurice M, *The Amazing Lid of Palenque I & II*, Brooks Hill Perry & Co., 1994.

Cotterell, Maurice M, *The Tutankhamun Prophecies*, Headline, London, 1999.

Covey, Curt, *The Earth's Orbit and the Ice Ages*, Scientific American,

David, Gary A, *The Orion Zone*, Adventures Unlimited Press, Kempton, IL. 2006

Dawood, N.J. *The Koran*, Penguin Classics, 1999.

DeVore & Sons, *Holy Bible*, Heirloom Bible Publishers, Wichita, Kansas, 1988.

Donnelly, Ignatius, *Atlantis The Antediluvian World*, Harper & Brothers, New York, 1882

Dutt, Romesh, C., *The Ramayana and Mahabharata*, Dover Publications, Inc., Mineola, New York, 2002.

Eliade, Mircea, *Patterns in Comparative Religion*, Meridian Books, Cleveland, 1966.

Erikson, Jon, *Ice Ages TAB Books*, Blue Ridge Summit, 1990.

Faulkner, O. Raymond, *The Egyptian Book of the Dead*, Chronicle Books, San Francisco, 1998.

Folger, Tim, *Waves of Destruction*, Discover, May 1994.

Gaer, Joseph, *How the Great Religions Began*, Dodd, Mead & Company, New York, 1957.

Gaddalla, Moustafa, *Tut-Ankh-Amen*. The Living Image of the Lord, Bastet Publishing, Erie, PA., 1997.

Gaspar, William, *The Celestial Clock*, Adam & Eva Publishing, Clovis, N.M., 2002.

Gaspar, William, *Sacred Cosmic Marriage*, A & E Publishing, Inc., Holman, N.M., 2005.

Geryl, Patrick and Ratinckx, Gino, *The Orion Prophecy*, Adventures Unlimited Press, Kempton, Illinois, 2001.

Gideons International, *The Holy Bible*, Thomas Nelson, Inc., 1985.

Gilbert, Adrian and Cotterell, Maurice, *The Mayan Prophecies*, Element, Boston, 1995.

Glatzmaier, Gary A, *The Geodynamo*. UCSC.

Glatzmaier, Gary A and Roberts, Paul, "*Rotation and Magnetism of Earth's inner core*" Science, 274, 1887-1891, 1996.

Guerber, H. A., *The Myths and Greece and Rome*, Dover Publications, Inc., Mineola, 1993.

Guthrie, Kenneth S, *The Pythagorean Sourcebook and Library*, Phanes Press, Grand Rapids, Michigan, 1988.

Halevi, Z'ev, ben Shimon, *The Work of the Kabbalist*, Samuel Weiser, Inc., Maine 1986.

Hall, Stan *Tayos Gold / The Archives of Atlantis*, Adventures Unlimited Press, Kempton, IL. 2006

Hancock, Graham, *Fingerprints of the Gods*, Crown Trade Paperbacks, New York, 1995.

Hathaway Dr., David H, *The Solar Dynamo*, NASA / Marshall Solar Physics.

Hoppal, Mihaly, *Sámánok*, Helikon Publishers, Budapest, 1994.

Jaramillo, José, *Between Two Worlds*, DM Presents, Digital Video Imaging

James, T.G.H., *Tutankhamun*, MetroBooks, New York, 2000.

James, T.G.H., *Ramesses II, White Star* / Friedman & Fairfax, New York, 2002.

Janson, F. Anthony, *History Of Art*, Prentice Hall Inc., New York, 1991.

Jenkins, M. John, *Maya Cosmogenesis 2012*, Bear & Co. Publ. Santa Fe, N.M. 1998.

Jenkins, M. John, *Galactic Alignment*, Inner Traditions Publ., Rochester, Vermont, 2002.

Johnsen, Berit, *The Cosmic Wedding*, by Frydenlund, Denmark,

Kerr, Richard A, *The Whole World Had a Case of the Ice-Age Shivers*, Science, 24, December, 1993.

Kersey, Graves, *The World's Sixteen Crucified Saviors*, Adventures Unlimited Press, Kempton, 2001.

Kovacs, Maureen Gallery, *The Epic Of Gilgamesh*, Stanford University Press, Stanford, 1989.

LaViolette, Paul, *Earth Under Fire*, Starlane Publications, Schenectady, 1997.

Lesko, Barbara, S., *The Great Goddesses of Egypt*, University of Oklahoma Press, Norman, 1999.

Levy, David H., *Skywatching*, Time Life Books, San Francisco, 1995.

Majorowski, Karen, *Mayan Sacred Calendar 2008*, Count of Days, Longmont, CO,

Mariner, Rodney, Rabbi, The *Torah*, Henry Holt and Company, Inc., New York, 1996.

Maspero, Gaston Sir, *Popular Stories Of Ancient Egypt*, Oxford University Press, New York, 2004.

Miller, Brent, *The Horizon Project*, DVD, Amazon.com

Muller, W. Max, Egyptian *Mythology*, Dover Publications, Inc., Mineola, N.Y. 2004.

Newcomb's, F.J., *Navaho Folk Tales*, University of New Mexico Press, 1991.

NOAA Paleoclimatology Program, *Educational Outreach*, Milankovitch Theory

O'Flaherty, Doniger Wendy, *Hindu Myths*, Penguin Books, London, 1975.

O'Flaherty, Doniger Wendy, *The Rig Veda*, Penguin Classics, 1981.

Ovid, *The Metamorphoses*, P.W. Suttaby and B. Crosby & Co., London, 1807.

Patrick, Richard and Croft, Peter, Classic *Ancient Mythology*, Crescent Books, New York, 1988.

Petrie, Flinders, W.M., *Egyptian Tales*, Dover Publications, Inc., Mineola, N.Y. 1999.

Pike, Albert, *Morals And Dogma*, AASR / TSCOTTSJ, Charleston, 1871.

Plato, *Timaeus*, Liberal Arts Press, New York, 1959.

Potok, Chaim, *Wanderings: History of the Jews*, Fawcett Crest, New York, 1980.

Raymo, Chet, *365 Starry Nights*, Fireside / Simon & Schuster, Inc., New York, 1982.

Roberts, T.R. & Roberts, M.J. and Katz, B.P., *Mythology*, MetroBooks, New York, 1997.

Ruddiman, W.F. and McIntire, A. *Oceanic Mechanisms For Amplifications Of The 23,000 Year Ice Volume Cycle*, Science, Vol. 212, # 4495, 8 May 1981.

Rukl, A, *The Encyclopedia Of Stars & Planets*, Ivy Leaf Publishing, London, 1982.

Santillana De, Giorgio and Von Dechend, Hertha, *Hamlet's Mill*, Gambit, Boston, 1969.

Schumann-Antelme, Ruth and Rossini, Stephane, *Illustrated Hieroglyphic Handbook*, Sterling Publishing Co., Inc., New York, 2002.

Schwaller de Lubicz, R.A. *The Temple Of Man*, Inner Traditions International, Rochester, Vermont, 1998.

Standing Bear, Luther, *My People the Sioux*, University of Nebraska Press, Lincoln, 1975.

Standing Bear, Luther, *My Indian Boyhood*, University of Nebraska Press, Lincoln, 1988.

Stray, Geoff, *Beyond 2012*, Vital Signs Publishing, East Sussex, 2005.

Seredy, Kate, *The White Stag*, Viking Press, New York, 1969.

Shearer, Tony, *Lord of the Dawn*, Naturegraph Publishers, Happy Camp, CA, 1995.

Suckling, Nigel, *Year of the Dragon / Legends and Lore*, Barnes & Noble, New York, 2000.

Tedlock, Dennis, *Popol Vuh*, Touchstone, New York, 1996.

Ulansey, David, *The Origins Of The Mithraic Mysteries*, Oxford University Press, New York, 1989.

Waters. Frank, *Book Of The Hopi*, The Viking Press, New York, 1970.

West, John Anthony, *Serpent In The Sky*, Versa Press Theosophical Publishing House, Wheaton, 1993.

Wilkinson, Richard, H. *Reading Egyptian Art*, Thames and Hudson, London, 1992.

Breinigsville, PA USA
10 November 2010
249064BV00003B/135/P